From Archangel to Senior Crown:
Design and Development
of the Blackbird

From Archangel to Senior Crown: Design and Development of the Blackbird

Peter W. Merlin
TYBRIN Corporation
History Office, NASA Dryden Flight Research Center
Edwards, California

Ned Allen, Editor-in-Chief
Lockheed Martin Corporation
Palmdale, California

Published by
American Institute of Aeronautics and Astronautics, Inc.
1801 Alexander Bell Drive, Reston, VA 20191-4344

American Institute of Aeronautics and Astronautics, Inc., Reston, Virginia

1 2 3 4 5

Library of Congress Cataloging-in-Publication Data

Merlin, Peter W., 1964-
 From archangel to senior crown : design and development of the Blackbird / Peter W. Merlin ; Ned Allen, editor-in chief.
   p. cm.
 Includes bibliographical references and index.
 ISBN 978-1-56347-933-5 (alk. paper)
 1. SR-71 Blackbird (Jet reconnaissance plane)--History. 2. A-12 Blackbird (Jet reconnaissance plane)--History. 3. Aerodynamics, Transonic--Research--United States--History--20th century. I. Allen, Ned. II. Title.

UG1242.R4M467 2008
623.74'67--dc22
2008000984

Cover design by Virginia Kozlowski

Copyright © 2008 by the American Institute of Aeronautics and Astronautics, Inc. The U.S. Government has a royalty-free license to exercise all rights under the copyright claimed herein for govermental purposes. All other rights are reserved by the copyright owner. Printed in the United States of America. No part of this publication may be reproduced, distributed, or transmitted, in any form or by any means, or stored in a database or retrieval system, without the prior written permission of the publisher.

FOREWORD

The Library of Flight is part of the growing portfolio of information services from the American Institute of Aeronautics and Astronautics. It augments the two established book series of the Institute—the Progress in Astronautics and Aeronautics Series and the AIAA Education Series—with the best of a growing variety of other topics in aerospace from aviation policy, to case studies, to studies of aerospace law, management, and beyond. Peter W. Merlin, one of the careful historians in NASA's Dryden History Office, here tells the story of the SR-71 Blackbird, one of the greatest aviation achievements of all time. In conceiving and realizing it, the Lockheed Skunk Works reached decades into the future for technologies that would not be matched for 50 years. The "SR" was a workhorse of Western strategic intelligence until that role was passed to overhead satellites, and it may have achieved the ultimate for what a turbine and ramjet manned platform could do—for no substantial element of its performance was matched until the first scramjet flight tests of the 21st century.

The Library of Flight seeks to document the crucial role of aerospace in enabling, facilitating, and accelerating global commerce, communication, and defense. Distinct from the Institute's other series, the lively Library of Flight authors often express opinions on matters of policy and controversy. As new aerospace programs grow and change around the world, we plan for the Library to host a wide array of international authors, expressing their own points of view on aerospace visions, events, and issues. As the demands on the world's aerospace systems grow to support new capabilities like unmanned vehicles, international relief, agricultural management, environmental monitoring, and others, the series will seek to document the landmark events, emerging trends, and new views.

Ned Allen
Editor-in-Chief
Library of Flight

TABLE OF CONTENTS

PREFACE ix
ACKNOWLEDGMENTS. xi

CHAPTER 1 BLACKBIRD: A TECHNOLOGICAL CASE STUDY 1
Note to Readers 4

CHAPTER 2 FORM FOLLOWS FUNCTION: DESIGN EVOLUTION
OF THE BLACKBIRD. 5
Convair Entries 7
Kelly's Archangel 8
A-12 Configuration 19
Aircraft Systems 23
OXCART—Progenitor of the Blackbird Family Tree. 31
KEDLOCK—Advanced Interceptor. 36
TAGBOARD—Mach 3 Drone 42
SENIOR CROWN—Strategic Reconnaissance 52

CHAPTER 3 NECESSITY IS THE MOTHER OF INVENTION:
CONSTRUCTION AND MATERIALS. 57
Building the Blackbirds. 57
Materials and Structure. 65
Exotic Materials. 68
Structural Features 70
Engine Materials 80
Fuels, Fluids, and Sealants. 82
Landing Gear 84

CHAPTER 4 ABOVE AND BEYOND: BLACKBIRD PERFORMANCE
CHARACTERISTICS. 87
Aerodynamics. 87
Handling Qualities. 89
Flight Control System 89
Normal Operating Characteristics. 92
Propulsion System 95
Speed and Altitude. 105

Cruise and Climb Performance *110*
Range and Endurance *111*
Payload Capacity and Operational Considerations *113*

CHAPTER 5   A UNIQUE RESEARCH TOOL: NASA's MACH 3
FLYING LABORATORY **115**
Joint NASA/U.S. Air Force YF-12 Flight Research *116*
Heating and Loads Research *119*
Propulsion Research *121*
Landing Studies *124*
Flying Laboratory *129*
Back in Black *139*
Phoenix Rising *142*
Swan Song *143*

CHAPTER 6   LESSONS LEARNED **147**
Kelly's Way—The Skunk Works Approach *147*
An Enormous Technological Achievement *154*
Results of the NASA Blackbird Research Programs *160*

APPENDIX A: BLACKBIRD PRODUCTION SUMMARY **167**
APPENDIX B: BLACKBIRD TIMELINE **177**
NOTES **183**
INDEX **191**
SUPPORTING MATERIALS **199**

Preface

I have been fascinated by the Lockheed Blackbirds for many years. Originated in secrecy, built of exotic materials, flown on critical intelligence-gathering missions around the globe, and painted a sinister matte black, these airplanes captured my imagination at an early age. Although the first A-12 flew two years before I was born and the last SR-71 made its last flight in 1999, it still bears the appearance of a futuristic design concept.

Not surprisingly, there have been numerous published books and articles about the Blackbirds. They have been primarily concerned with the operational history of the airplane. I, however, chose to concentrate on the developmental history of the airplane's design, technological and performance aspects, use of the aircraft as a research platform, and lessons learned form the unique challenges of designing, building, and flying these fabulous planes. Although this study is technical in nature, it is also intended to be accessible to a general readership and should be of value to historians, teachers, aeronautical engineering students, and aviation buffs.

Peter W. Merlin
December 2007

ACKNOWLEDGMENTS

The author would like to thank the many people who made this project possible. A large number of individuals and organizations provided invaluable support and suggestions. First of all, thanks to Tony Springer, NASA Aeronautics Research Mission Directorate, for suggesting and sponsoring this project. I am grateful for the efforts of many people at NASA Dryden Flight Research Center including, but not limited to, Everlyn Cruciani, Christian Gelzer, Jay Levine, Steve Lighthill, Jerry McKee, Sarah Merlin, Steve Parcel, and Mike Relja. Special thanks to a number of personnel at Lockheed Martin Aeronautics Company, especially communications director Dianne Knippel who kept a variety of complex processes running smoothly in order to facilitate completion of the book, and Bill Simone for efforts above and beyond the call of duty. Pete Perez and George Soto reviewed my material for technical accuracy. Thanks to the staff of the Air Force Flight Test Center, particularly Raymond Puffer and Stephanie Smith, for providing valuable source material. Thanks to Gene Matranga, formerly of NASA and Lockheed, for participating in oral history interviews and providing material from his personal files. I am grateful to Jim Goodall, author and self-appointed Blackbird historian, for providing many amazing photos. Thanks to Paul Kucher for taking the time to meticulously scan many pages of technical documentation. Special thanks to the membership of Roadrunners Internationale who graciously allowed me into their ranks as an associate member so that I could record their history for posterity. Apologies to anyone I missed. Any factual errors are the author's responsibility. I made an attempt in good faith to get the facts straight by using the best available source material.

Chapter 1

BLACKBIRD: A TECHNOLOGICAL CASE STUDY

The Lockheed Blackbirds hold a unique place in the development of aeronautics. In their day, they outperformed all other jet airplanes in terms of altitude and speed. Now retired, the Blackbirds remain the only production aircraft capable of sustained Mach 3 cruise and operational altitudes above 80,000 feet.

Conceived as airborne reconnaissance platforms, the family of aircraft known collectively as the Blackbirds included the A-12, YF-12, M-21, and SR-71. Designed by Clarence L. "Kelly" Johnson under the nickname "Archangel," the A-12 resulted from a series of designs for a successor to Lockheed's earlier U-2 spy plane. The twelfth design in Johnson's Archangel series was a sleek aircraft built almost entirely of titanium. With powerful turboramjets, the A-12 was capable of attaining a cruise speed of Mach 3.2 and an operational altitude of 90,000 feet.

In August 1959, the Central Intelligence Agency (CIA) approved funding for construction of the A-12 as Project OXCART. Between 1960 and 1962, Lockheed engineers tested a scale model of the A-12 in NASA Ames Research Center's 8 foot × 7 foot Unitary Plan High-Speed Wind-Tunnel at Moffett Field, California. These tests included various inlet designs, control of cowl bleed, design performance at Mach 3.2, and off-design performance of an optimum configuration up to Mach 3.5.

As a reconnaissance platform, the A-12 was flown exclusively by the Central Intelligence Agency. The first airframe was delivered in February 1962 and made its maiden flight in April. Test flights and operational missions continued until June 1968. A-12 pilots required full pressure suits, enabling them to fly for extended periods at Mach 3.2 and at operational altitudes of 70,000 to 90,000 feet.

In March 1960, even before delivery of the first A-12 prototype, Lockheed and the Air Force discussed development of an interceptor version of the A-12. Designed as the AF-12 under Project KEDLOCK, the interceptor featured a pulse-Doppler radar system and launch bays for three air-to-air missiles. A second crew position, located just behind the cockpit, accommodated a fire control officer (FCO) to operate the missile launch system. With

Lockheed engineers tested a scale model of the A-12 in the 8 × 7 foot Unitary Plan Wind Tunnel at NASA Ames Research Center, Moffett Field, California, in the early 1960s. (Courtesy of Lockheed Martin Corp.)

the assistance of the CIA, the Air Force entered into an agreement with Lockheed to build three prototypes, eventually designated YF-12A, and the maiden flight took place in August 1963.

The public first became aware of the aircraft on February 29, 1964, when President Lyndon B. Johnson announced its existence. By agreement with Kelly Johnson, the president intentionally misidentified the aircraft as an "A-11." Now public knowledge, the YF-12A flight-test program was moved

The Lockheed A-12 was designed as a reconnaissance platform capable of cruising at Mach 3 speeds and altitudes between 70,000 and 90,000 feet. (Courtesy of Lockheed Martin Corp.)

to Edwards Air Force Base, northeast of Los Angeles in the Mojave Desert. The Air Force soon began testing the aircraft's weapons system and worked on resolving troublesome issues with transonic acceleration and various subsystems.

The year 1964 also marked the debut of two more Blackbird variants, designated M-21 and SR-71. The M-21, a two-seat variant of the A-12, was built expressly as a launch aircraft for the secret D-21 reconnaissance drone. A fatal accident during the fourth launch resulted in destruction of both drone and M-21 and the death of the launch control officer. The second new Blackbird, the SR-71, became the most familiar member of the family. Operated by the U.S. Air Force under Project SENIOR CROWN, the SR-71 served as an aerial reconnaissance workhorse around the world for more than 25 years.

The A-12 fleet operated in secret for six years, during which time several were lost in crashes. By June 1968 all surviving airframes had ended their service lives and been sent to Lockheed's facility in Palmdale, California, for permanent storage. Their operational mission had been assumed by the SR-71A, operated by the Air Force. A planned operational version of the YF-12A interceptor, designated F-12B, failed to materialize as Secretary of Defense Robert McNamara ultimately cancelled the program as a cost-cutting measure. As a consequence, on December 29, 1967, Air Force officials instructed Lockheed to terminate F-12B development. The YF-12A program ended in February 1968, and the aircraft joined the A-12 fleet in storage. There they remained until the National Aeronautics and Space Administration (NASA) reached an agreement with the Air Force for a joint research program. Beginning in 1969, NASA operated two YF-12A aircraft and one SR-71A (temporarily designated YF-12C for political reasons). The joint NASA–Air Force program continued for 10 years.

The A-12 design evolved into the SR-71, a multisensor platform capable of carrying a wide variety of reconnaissance equipment. (Courtesy of NASA)

The Air Force retired the SR-71 fleet in 1990, but two airframes were reactivated for operational service in 1995. They were retired again in 1997. NASA operated the SR-71 between July 1991 and October 1999 for research purposes and to support the Air Force reactivation program. After retirement from NASA service, all remaining Blackbird airframes were allocated to museums and former operating agencies for permanent display.

Over the years, numerous books and articles have been written about the Blackbirds. Previous authors have provided brief overviews of the technological aspects while concentrating on the developmental and operational history of these incredible airplanes. This monograph will explore the technological aspects of the Blackbird family and the lessons learned through the process of designing, building, and operating them.

NOTE TO READERS

The various organizations that worked with the Blackbirds referred to individual airframes by numerical designations. Lockheed technicians and designers used what were known as "article numbers." The first A-12 was called "Article 121." The A-12 production run included articles 121 through 133. The two M-21 airframes were articles 134 and 135. The three YF-12A interceptors were articles 1001, 1002, and 1003. D-21 drones included articles 501 through 538. The SR-71 airframes were known as articles 2001 through 2031. Part of structural test article 2000 was later incorporated into a flyable aircraft (SR-71C) and retained its number. Air Force personnel used Department of Defense serial numbers allocated according to fiscal year of acquisition. A-12 serials include 60-6924 through 60-6933 and 60-6937 through 60-6939. The YF-12A airframes were assigned serial numbers 60-6934, 60-6935, and 60-6936. The M-21 variants were numbered 60-6940 and 60-6941. D-21 drones were never assigned military serials. SR-71 serials included 61-7950 through 61-7981. In the past, SR-71 serial numbers have been published erroneously with a "64" fiscal-year prefix (such as 64-17950), but no such prefix was ever used. All official documentation bears this out. NASA assigned three-digit numbers to each aircraft in its inventory. For the YF-12 research program, the last three digits of the tail number became the NASA number. In the later SR-71 program, each aircraft received a unique 800-series number identifying the airplane's affiliation with NASA Dryden Flight Research Center. For the purposes of the text, the author will usually use the Lockheed article numbers to identify specific airframes. All numerical designators are cross-referenced in Appendix A.

Chapter 2

FORM FOLLOWS FUNCTION: DESIGN EVOLUTION OF THE BLACKBIRD

Development of the family of aircraft known collectively as the Lockheed Blackbirds began with a requirement for a successor to the U-2. A reconnaissance airplane capable of high-altitude (but low-speed) flight, the U-2 made its first flight in 1955. It was built by Lockheed's Advanced Development Projects Division (known as the "Skunk Works") under the Central Intelligence Agency's (CIA) Project AQUATONE, directed by Chief Engineer Kelly Johnson.

CIA and Air Force analysts concluded the U-2 would have a relatively short operational life span before hostile antiaircraft technology rendered it obsolete. As early as 1956, at the same time the U-2 was becoming operational, Johnson proposed a Mach 2.5 hydrogen-fueled airplane capable of

The Lockheed U-2 was capable of flight in excess of 70,000 feet, but was subsonic and easily detectable by radar. It was thought that improvements in hostile antiaircraft technology would render the U-2 obsolete within a few years of attaining operational capability. (Courtesy of Lockheed Martin Corp.)

cruising above 99,000 feet. Only 25 people were cleared into this special access program, code named SUNTAN.[1]

Lockheed's initial SUNTAN studies evolved into a vehicle nearly 300 feet long with a gross takeoff weight of 358,500 pounds. Johnson's CL-400 design posed numerous technical challenges involving materials, manufacturing, airframe/powerplant integration, fuel production, and storage and handling. As technological problems began to overwhelm the project, Johnson began to have serious doubts about its viability.

After further analysis of the SUNTAN design and proposed mission requirements, he determined the CL-400 had severe range limitations that could not be overcome with available technology. During a 1957 meeting with the secretary of the Air Force, Johnson recommended terminating the program in favor of one focused on an airplane with a more conventional propulsion system. He advocated a smaller, lighter airframe powered by two Pratt and Whitney J58 engines. In February 1959, SUNTAN was finally terminated at Johnson's request, and his final design, the hydro-carbon-fueled CL-400-15JP, served as a stepping stone to the Archangel project that eventually yielded the A-12.[2]

Lockheed designers had relied primarily on altitude to protect the earlier U-2 from hostile ground fire, but they underestimated the ease with which the airplane could be tracked by radar. Attempts to reduce its radar cross section (RCS) using radar-absorbent coatings and structures were not very effective. Such treatments also resulted in significant weight and performance penalties and, in one instance, contributed to the loss of a test aircraft and pilot.[3]

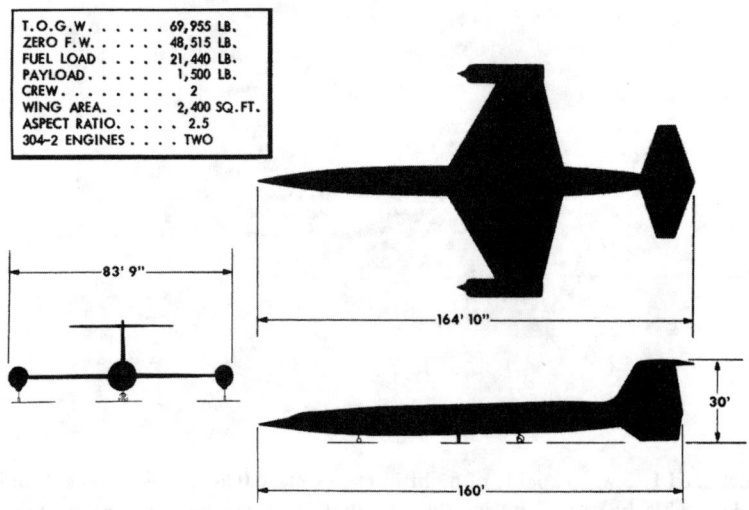

Lockheed's Hydrogen-fueled CL-400 was designed to cruise at altitudes in the vicinity of 99,000 feet. Insurmountable technical difficulties led to termination of the project before any airframes were built. (Courtesy of Lockheed Martin Corp.)

In the fall of 1957, the CIA commissioned a study to determine the probability of detecting an airplane by radar with respect to its speed, altitude, and RCS. This analysis indicated that supersonic speed significantly reduced the ability of conventional radar systems to detect an aircraft.

Most tracking radars at the time swept a 30- to 45-degree band of the sky. Reflected radar pulses showed up on monitor screens as a spot or blip, proportional to the size of the object. The persistence of this blip on the screen was dependent on the number of times a target signal was obtained divided by the number of antenna scans over a given period. This blip/scan ratio (BSR) was affected by the strength of the radar return, the altitude of the object being tracked, and its airspeed. Results of the CIA study indicated that an airplane with a radar cross section of less than 10 square meters, moving at high speed at altitudes approaching 90,000 feet, would be extremely difficult to track because of its low BSR.[4]

Subsequently the CIA, under Project GUSTO, solicited design proposals from Lockheed and the Convair Division of General Dynamics. The U.S. Navy also submitted an in-house concept, but it had insurmountable design flaws.

The Navy proposed a ramjet-powered, inflatable rubber aircraft that would climb to altitude beneath a balloon. A rocket would then propel the craft to speeds at which its ramjet engine could function. Unfortunately, studies determined the balloon would need a diameter of one mile in order to lift the craft to altitude. The airplane itself required a wing area greater than one-seventh of an acre.[5]

CONVAIR ENTRIES

Lockheed's most serious competition came from Convair's proposed ramjet-powered vehicle, codenamed FISH, which was to be carried aloft beneath a B-58 supersonic bomber. To survive extreme aerodynamic heating and also have a minimal radar signature, the vehicle was to be constructed using Pyro-Ceram (a ceramic glass having virtually zero thermal expansion under extreme heating conditions) and other heat-resistant, radar-attenuating materials. Following a Mach 2.0 launch from a lengthened B-58 with uprated engines, two Marquardt ramjets would propel FISH during the Mach 4.25 cruise portion of its mission. The pilot would then deploy two pop-out General Electric J85 turbojet engines for maneuvering during the landing phase. The Convair team, headed by Robert Widmer and Vincent Dolson, concluded the vehicle's size, propulsion system, and logistics were impractical. As well, it would have been difficult to support in an operational environment. Subsequently, they scrapped FISH and began to formulate a concept of a vehicle that could function autonomously, without the need for a launch aircraft.

Built around two Pratt and Whitney JTD11D-20 (J58) engines, the new craft was called KINGFISH. It was larger than FISH and capable of taking

"FIRST INVISIBLE SUPER HUSTLER"
1958-1959

Cruise Mach Number: 4.2
Cruise Altitude: 90,000 ft
Range: 3,900 NM
Span: 37.0 ft
Length: 48.5 ft
Height: 9.8 ft

Propulsion:
Two "pop-out" J85 turbojets for landing
Two Marquardt ramjets for cruise

Launched from B-58 at Mach 2.2 above 35,000 ft

Convair's ramjet-powered FISH proposal was a serious contender in the race to design a successor to the U-2. Its major disadvantage was the requirement that it be launched from another airplane. (Courtesy of John R. Whittenbury via Lockheed Martin)

off under its own power, but would have a top cruise speed of approximately Mach 3.25 at an altitude of 125,000 feet. A much larger vehicle, KINGFISH was designed to carry two crew members and a large sensor package. The vehicle's wing edges were to be built in a complex pattern of interlocking wedges, every other one made of radar-absorbent material to reduce the radar cross section (RCS). Convair built a model of the KINGFISH airframe for radar signature tests, but never produced a flyable airframe.

Meanwhile Lockheed struggled to produce a viable design. Tentatively called the U-3 in early Skunk Works studies, the airplane had to meet stringent RCS requirements to make it more survivable than the U-2 to hostile anti-aircraft defenses. Kelly Johnson developed and discarded numerous designs in an attempt to meet the Central Intelligence Agency's specifications. Although he could design an airplane capable of attaining high speeds and altitudes, he found it difficult to significantly reduce the radar signature. For a while it looked as if the contract would go to General Dynamics/Convair.

By August 1959 Johnson had offered the CIA a total of 11 high-speed proposals and two low-speed, low-RCS concepts. Just when it looked as if KINGFISH would be the winner, the CIA agreed to accept an airplane with a lower cruising-altitude capability if it had the desired radar cross section and speed. Johnson was able to modify his design and the 12th Lockheed concept was selected for production.[6]

KELLY'S ARCHANGEL

Johnson's Archangel concepts, numbered A-1 through A-11, were driven by the need for speed and altitude, but customer requirements for survivability ultimately led to a revolutionary design with a small radar signature. Along the path that eventually led to the A-12, Johnson explored an eclectic collection of alternative design concepts.

His first rough pencil sketch on April 23, 1958, for a Mach 3.0 airplane (then still called U-3), featured a slender, tapered airframe with a cross section that was cylindrical up to the point where two engine pods nestled tightly against the aft fuselage. The high-mounted wing featured a diamond shape with squared tips. Two widely spaced, outwardly canted vertical stabilizers and two variable-position horizontal surfaces provided longitudinal and lateral control. At this point the CIA's stated design objectives included a 500-pound reconnaissance payload capability, unrefueled mission radius of 2000 nautical miles, and 90,000-foot cruising altitude.[7]

To make his paper airplane a reality, Johnson and his team of engineers had to draw on available enabling technologies and design tools. Routine calculations were made using slide rules. More complex calculations, such as stress analysis, required Friden mechanical calculators. The most advanced computer available at the time was the IBM mainframe. With the era of computational fluid dynamics still far in the future, aerodynamic and loads testing was limited to that performed using wind-tunnel models.

Balancing the need for a lightweight structure against resistance to aerodynamic heating, the Skunk Works team focused on B120-VCA titanium

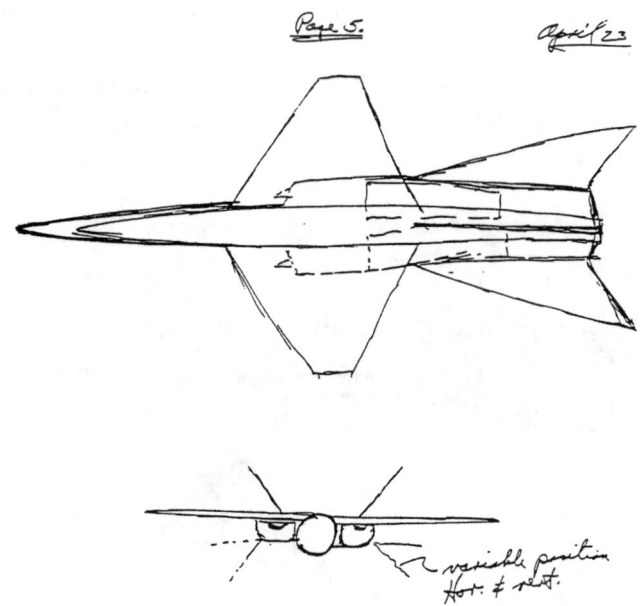

Clarence "Kelly" Johnson completed his first pencil sketch of what was then known as the U-3 on April 23, 1958. It was intended to carry a 500-pound payload to 90,000 feet altitude. (Courtesy of Lockheed Martin Corp.)

alloy as the primary structural material. For propulsion, Pratt and Whitney's J58 turbojet engine offered the best performance at high-Mach cruise conditions while mixed-compression, variable-geometry inlets maximized inlet recovery across the flight envelope. Engineers initially felt separate ramjets might be used to enhance high-altitude performance at cruise Mach numbers. The most promising available fuel was JP-150, with low vapor pressure and stability over a wide temperature range. High-energy fuels such as pentaborane and ethyldecaborane were considered and rejected. They offered a 35% increase in energy content per unit mass vs JP-150, but because of their high toxicity presented logistical difficulties.[8]

By July 1958 the U-3 had evolved into Archangel 1. The earliest version of the A-1 design featured a conventional fuselage, 166.67 feet long, with a sharply pointed nose. The wing, spanning 49.6 feet, was mounted at the top of the airframe and featured a sharply swept leading edge and gently swept trailing edge. Two J58 engines occupied engine pods below the wing roots. The cruciform tail assembly featured conventional vertical and horizontal surfaces with a relatively large surface area.

The aircraft's empty weight was estimated at 41,000 pounds. With a capacity for storing 61,000 pounds of fuel, it would have had a gross weight of 102,000 pounds at takeoff. The A-1 was designed to cruise at a speed of Mach 3.0 and altitudes between 83,000 and 93,000 feet with a mission radius of 2000 nautical miles. Later versions of the A-1 explored such options as canards mounted on the forward fuselage, a double-delta wing configuration, and winglets.[9]

Even as Johnson struggled to refine his high-speed concept, he offered the CIA a low-speed vehicle called the G-2A. This design offered a tailless,

Archangel 1, or A-1, featured a fairly conventional configuration. In later iterations, the designer explored the possibility of using canards, a double-delta wing, and winglets. (Courtesy of Lockheed Martin Corp.)

flying wing with vertical surfaces near the tips. A stubby, blended body in the center provided space for a cockpit, sensor bay, and two turbojet engines buried in the structure. Its greatest advantage was an extremely small radar cross section. Johnson presented it to the CIA, where it was viewed as a potential backup to the high-performance vehicle, but the G-2A presented some unique problems. To reduce RCS, the airframe was to be constructed from a plastic material that was nearly invisible to radar. Unfortunately, when a model was tested in Lockheed's radar anechoic chamber, the internal equipment and fuel presented a larger RCS than an all-metal airplane of the same design called the G-2S.[10]

In August and September, Johnson worked on several versions of the A-2. These included a four-engine vehicle with two J58 turbojets beneath the wing at midspan to provide bending relief and ethyldecaborane-fueled, 75-inch-diameter ramjets at each wing tip. In addition the wing sweep angle was substantially reduced and the cruciform tail enlarged. With a length of 129.17 feet and a wing span of 76.68 feet, the A-2 was expected to cruise at Mach 3.2 and altitudes of 94,000 to 105,000 feet. It was projected to have a gross takeoff weight of 135,000 pounds, assuming a fuel capacity of 81,000 pounds. The ramjets would be ignited at Mach 0.95 and 36,000 feet altitude.

Both the A-1 and A-2 were rejected as a result of excessive gross weight and the still-unaddressed the RCS issue. Johnson responded by designing a scaled-down vehicle with smaller, modified Pratt and Whitney JT-12 engines.

The G-2A concept explored the possibility of building a subsonic aircraft with a very low radar cross section. Kelly Johnson viewed it as an alternative to a high-performance vehicle. (Courtesy of Lockheed Martin Corp.)

Because the JT-12 was a low-pressure-ratio engine, it was well suited to high-Mach-number operation.[11]

The A-3 had two JT-12 powerplants mounted at midspan with the engine nacelle centered in the wing structure. These turbojets looked insignificant compared to the 40-inch-diameter ramjets on the wing tips. The turbojets would provide thrust for takeoff, climb, and acceleration whereas the ramjets would be used only during the cruise portion of the flight. The A-3 of November 1958 was Johnson's smallest design to date, just over 62 feet long with a wing span of 33.8 feet. It had a gross takeoff weight of just 34,600 pounds and was expected to meet all performance requirements. A semi-tailless configuration (no horizontal stabilizer) reduced both weight and radar cross section.

In December Lockheed officials were asked to evaluate designs for a vehicle to be launched from a B-58 for comparison with the Convair FISH concept. The Skunk Works dutifully provided two delta-wing designs, each powered by two 40-inch-diameter ramjets, with a design cruise speed of Mach 4.0 above 95,000 feet. An additional JT-12 turbojet would provide extra thrust to help reduce the approach angle during landing. These designs, known as Johnson's Arrow series, had higher performance and lower RCS than previous concepts but gave rise to several critical concerns. These included reliability of the ramjet during cruise, lack of go-around capability

ARCHANGEL 2
SEPTEMBER 1958

Length:	129.17 ft	Zero Fuel Weight:	54,000 lbs	Cruise Mach:	3.2
Span:	76.68 ft	Fuel Weight:	81,000 lbs	Cruise Alt:	94 -105 kft
Height:	27.92 ft	Takeoff Gross:	135,000 lbs	Radius:	2,000 NM

75" dia ramjets burning HEF
(Lit @ Mach 0.95, 36,000 ft)

Reduced wing sweep compared to A-1

Two J58 turbojets with AB burning JP-150
(Moved further outboard for bending relief)

Kelly Johnson worked on several versions of the A-2, a four-engine vehicle with two fuselage-mounted turbojets and two wingtip-mounted ramjets. (Courtesy of John Whittenbury via Lockheed Martin)

FORM FOLLOWS FUNCTION 13

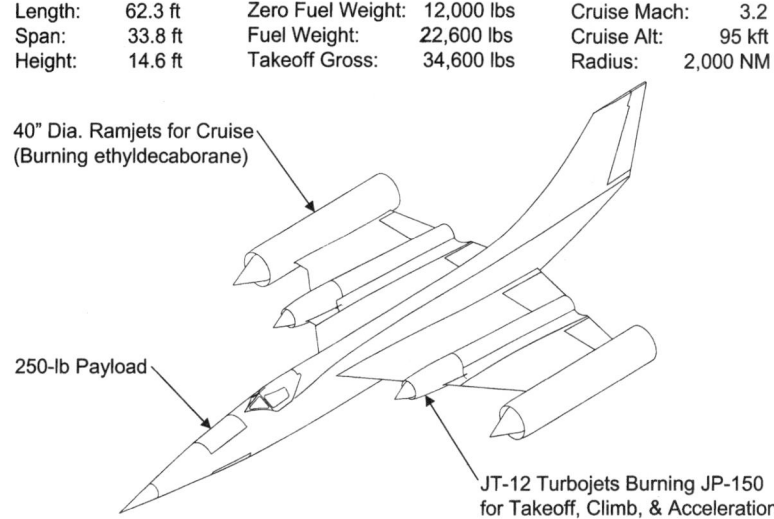

The semi-tailless configuration of the A-3 reduced weight and radar cross section. (Courtesy of John Whittenbury via Lockheed Martin)

during landing, minimal ground clearance while attached to the B-58 during takeoff, and no means of pilot ejection while the vehicle was attached to the launch aircraft.

In response to these problems, Johnson returned to his Archangel designs, which were capable of taking off under their own power. His next concept, the A-4, was relatively small at 58 feet long with a 35-foot span. Its blended wing-fuselage configuration significantly reduced the radar cross section. The vertical stabilizer resembled a shark's dorsal fin running the length of the upper fuselage. A single J58 served as the main powerplant while two 34-inch-diameter ramjets on the wing tips provided cruise power. Maximum gross takeoff weight was estimated at 57,900 pounds.

In an attempt to further reduce size and weight, Johnson proposed the A-5. At a length of 46 feet and span of 32.5 feet, it came in at a gross weight of just 50,320 pounds, but featured the most complex mix of powerplants yet. Two JT-12 turbojets, buried in blended side fairings, provided thrust for takeoff, climb, and landing. A centrally located 83-inch-diameter ramjet with a ventral intake provided cruise power while additional takeoff thrust came from a 10,000-pound-thrust liquid-fueled rocket at the base of the vertical fin. In all other respects the A-5 resembled a scaled-down A-4. Design integration was extremely challenging, particularly with respect to fuel accommodation.

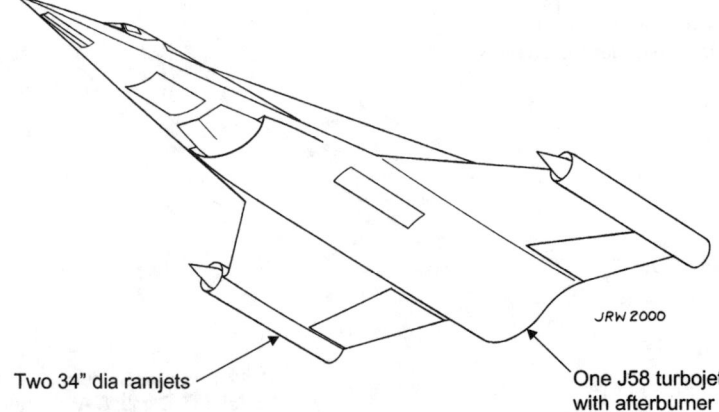

In another bid to reduce radar cross section, Johnson designed the A-4 with a blended wing and fuselage. (Courtesy of John Whittenbury via Lockheed Martin)

For the A-6, Johnson proposed a configuration with a blended triangular forebody and delta wings with squared tips. Inwardly canted vertical fins were located about two-thirds of the way out from the wing roots. The powerplants included a single J58 and two 34-inch-diameter ramjets buried in the fuselage. With a length of 64 feet and a span of 47 feet, the airplane had a gross weight of 62,950 pounds. Weight penalties were reduced by equipping the airplane with lightweight landing gear. A detachable set of heavy-duty gear would be used for takeoff and drop away as soon as the craft lifted off the ground.

By January 1959 a number of things had become clear. Maximum performance and minimum RCS seemed to be mutually exclusive. The A-4 through A-6 designs lacked the necessary operational range. Skunk Works engineers noted that ramjet technology was not sufficiently mature for use in long-range cruise conditions. Two-stage systems, such as those involving a B-58 launch aircraft, were operationally impractical for multiple reasons including logistics and safety. Additionally, the customer was understandably anxious by this point to see a finished product. Johnson began to focus on a maximum-performance turbojet aircraft design with no performance concessions for the sake of improved RCS.

This effort resulted in the A-7, a configuration similar to the A-1 and A-2 but scaled down to just under 98 feet long with a 47-foot wing span. It was

FORM FOLLOWS FUNCTION   15

A-5
DECEMBER 1958

Length:	46.0 ft	Zero Fuel Weight:	18,500 lbs	Cruise Mach:	3.2
Span:	32.5 ft	Fuel Weight:	31,820 lbs	Cruise Alt:	90 kft
Height:	16.92 ft	Takeoff Gross:	50,320 lbs	Radius:	1,557 NM

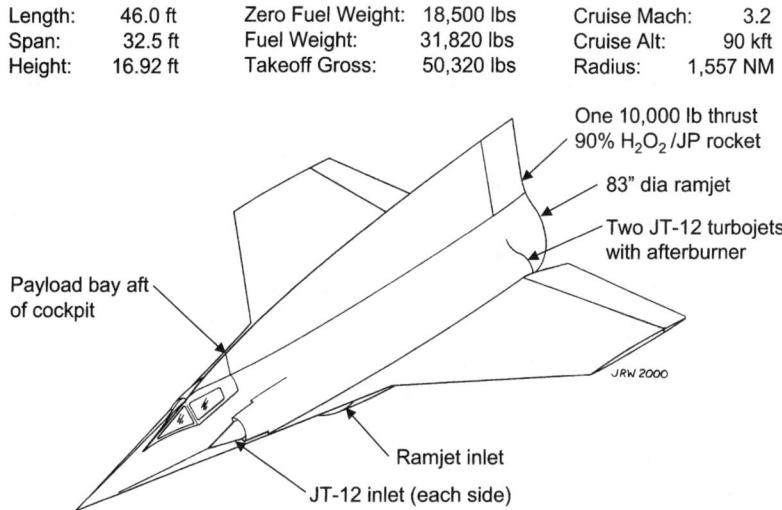

The A-5 had a complex mix of powerplants including two turbojets for low-speed flight, a single ramjet to provide cruise power, and a 10,000-pound-thrust liquid-fueled rocket for takeoff assistance. (Courtesy of John Whittenbury via Lockheed Martin)

A-6-5
JANUARY 1959

Length:	64.0 ft	Zero Fuel Weight:	29,200 lbs	Cruise Mach:	3.2
Span:	47.2 ft	Fuel Weight:	33,750 lbs	Cruise Alt:	90 kft
Height:	22.85 ft	Takeoff Gross:	62,950 lbs	Radius:	1,287 NM

Droppable gear for takeoff; lightweight gear for landing

For weight reduction, the A-6 was to be equipped with detachable heavy-duty gear for takeoff and permanently attached lightweight main gear for use during landing. (Courtesy of John Whittenbury via Lockheed Martin)

powered by a single J58 in the fuselage and two 34-inch-diameter Marquardt XPJ-59 ramjets on the wing tips. All engines would burn only JP-150 fuel. It had a projected maximum gross takeoff weight of 70,900 pounds.

Johnson continued to refine the concept with the A-8 and A-9 designs, but results were disappointing. Mission radius continued to hover around 1637 nautical miles with a cruise altitude of slightly more than 91,000 feet, considerably less than the A-2 despite the weight reduction.

In February 1959 Johnson submitted the A-10 concept, an elegantly simple design. The 109-foot-long cylindrical fuselage featured a long forebody and was sharply tapered at each end. The semi-double-delta wings had squared tips and spanned 46 feet. A vertical tail fin with conventional rudder provided lateral stability. Two General Electric J93-3 turbojets would propel the airplane to speeds of Mach 3.2 at a 90,000-foot cruise altitude. At a takeoff weight of 86,000 pounds, the A-10 demonstrated a significant weight reduction (18,000 pounds) relative to the A-1 and allowed it to reach higher altitudes. Mission radius was estimated at 2000 miles. Radar cross section was still a problem, but Johnson was more concerned with performance.

The following month he refined the design further. The A-11 featured true double-delta wings spanning 56.67 feet. Fuselage length increased to

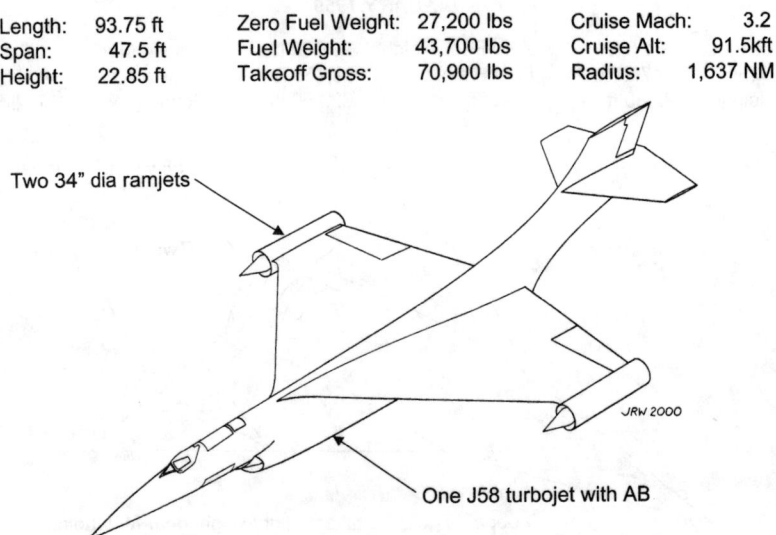

With the A-7, Johnson returned to a design similar to his original Archangel concepts. He continued to refine the concept with the A-8 and A-9, but results were disappointing. (Courtesy of John Whittenbury via Lockheed Martin)

Form Follows Function 17

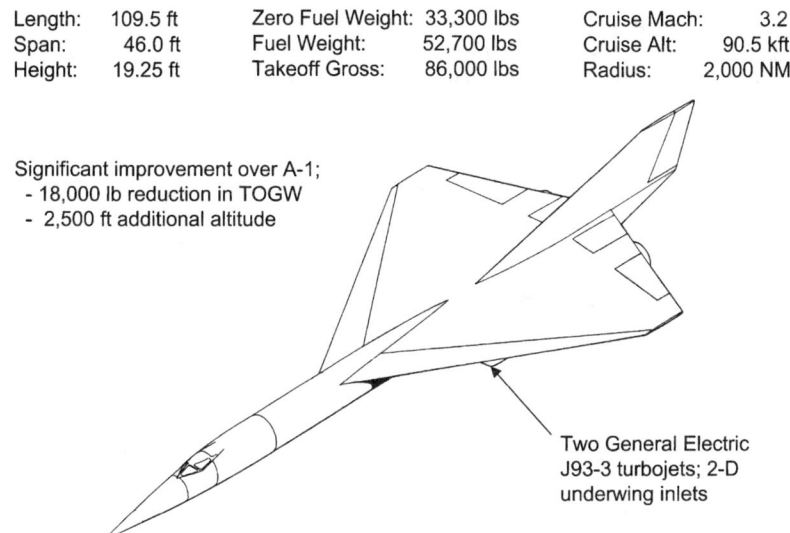

Johnson's A-10 design made it abundantly clear that he was more interested in performance than RCS reduction. (Courtesy of John Whittenbury via Lockheed Martin)

116.67 feet, and the J93 engines were replaced with J58s. The A-11 was designed to take off from a home base, cruise at Mach 3.2 at 93,500 feet, and complete an eight-hour, 13,340-nautical-mile mission with two aerial refuelings.[12]

Johnson pitched his A-11 concept to the CIA and reported the results of six months of radar studies. He emphasized that expected improvements to radar systems would enable detection of any airplane that might conceivably fly within the next three to five years. He specifically noted the probability of detection of the A-11 was practically 100%. It was subsequently agreed the airplane might make such a strong radar target that it could be mistaken for a bomber. This was unacceptable for an airplane that was intended for use in clandestine reconnaissance missions.

On July 3, 1959, the director of the Program Office at CIA headquarters paid Johnson a visit. Johnson thought Lockheed was about to be ruled out as a contender, but was relieved to learn instead that the CIA had offered to extend Lockheed's program and accept lower cruising altitudes in exchange for incorporation of RCS reduction techniques.

Johnson subsequently proposed the A-12 with the J58 engines in a midwing arrangement to reduce the airplane's side profile. Chines along the forebody reduced fuselage sloping while providing additional lift and stability. The single vertical stabilizer was replaced with two all-moving vertical fins,

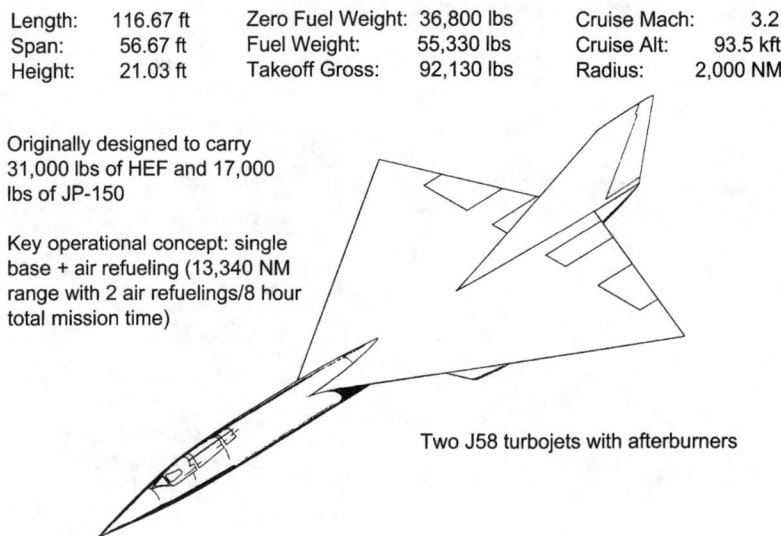

The A-11, with its double-delta wing was a truly elegant design. Despite promised performance capabilities, its radar cross section made the design unacceptable to the CIA. (Courtesy of John Whittenbury via Lockheed Martin)

one on top of each engine nacelle. These were canted inward for further RCS reduction. Serrations on the wing edges incorporated radar-absorbent materials. Johnson noted in his project diary, "This airplane weighs about 110,000 to 115,000 pounds and, by being optimistic on fuel consumption and drag, can do a pretty good mission."[13]

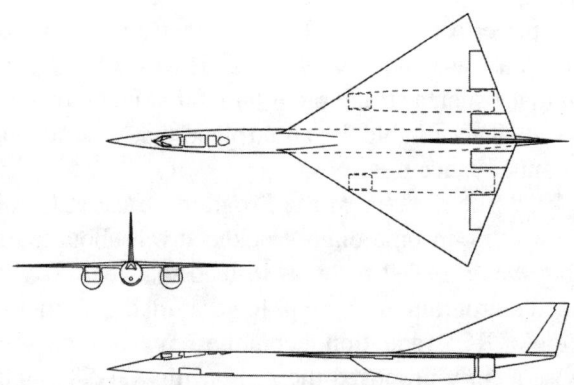

The A-11 was rejected because CIA offiicials believed that, because its radar return was so large, it might be mistaken for a bomber. (Courtesy of John Whittenbury via Lockheed Martin)

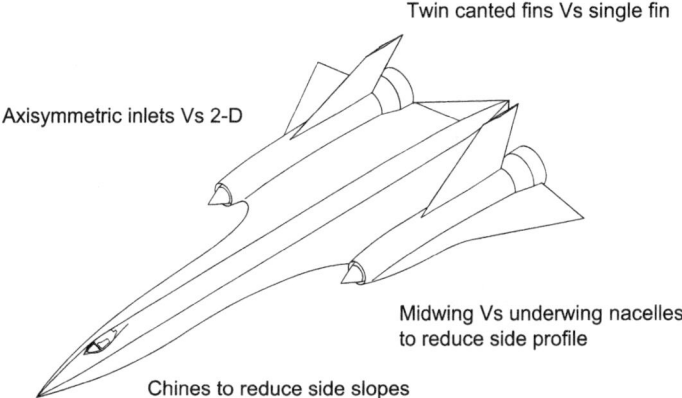

The initial A-12 configuration was Johnson's first attempt to design an airplane that combined low RCS with high performance. (Courtesy of John Whittenbury via Lockheed Martin)

Lockheed and Convair submitted final proposals to the CIA on August 20, 1959. Although Convair's KINGFISH promised better overall performance, the evaluation panel considered it to be technologically riskier. Subsequently, Lockheed was awarded a four-month initial contract with the admonition that it must prove the viability of its antiradar approach by January 1, 1960, before receiving full go-ahead. At this point Project GUSTO was terminated. The new airplane would be built under Project OXCART.[14]

A-12 CONFIGURATION

The A-12 design incorporated features to maximize performance, survivability, and mission capability while minimizing weight, detectability, and (to the extent possible) cost. The fuselage contained no wasted space or extraneous material, and even the fuel did double duty as a coolant. Johnson's Archangel redefined the state of the art in aeronautical design.

The airplane's fuselage consisted of a titanium structure of semimonocoque construction with a circular cross section. The sides flared out into sharply blended chines, assembled as interlocking saw-toothed wedges. On all A-12 airframes except those of the prototype and trainer variant, the outward-pointing teeth were fashioned from titanium while the interlocking, inward-pointing teeth were made from radar-absorbent composites.

The sharply tapered nose section was pressurized and contained navigational and communications equipment, a remote compass transmitter, periscope optics, air inlet computer and angle transducer, and other radio

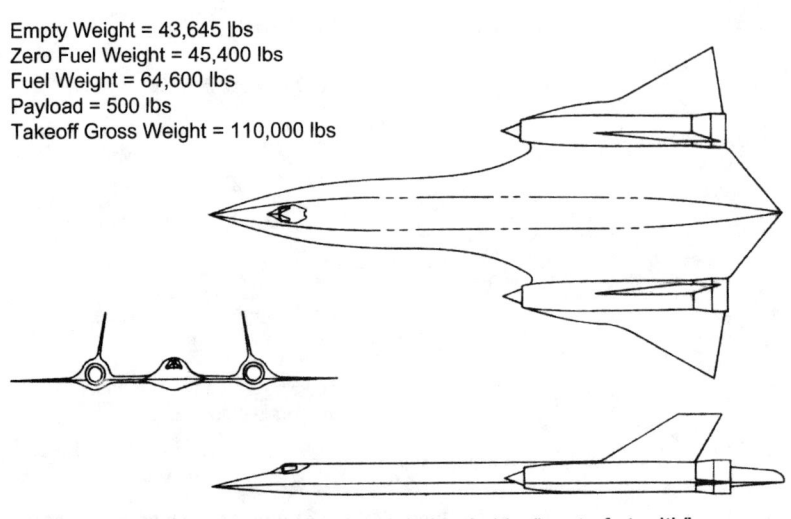

As Johnson refined his A-12 concept, the design gradually evolved into its final configuration. (Courtesy of John Whittenbury via Lockheed Martin)

equipment. A combination pitot-static and alpha-beta probe was installed at the forward tip to capture airspeed and altitude data.

The pilot's station (cockpit) featured a V-shaped windscreen and was enclosed by an aft-hinged clamshell canopy. Both the windscreen and canopy included windows with dual glass assemblies. The outer monolithic glass panels were separated from inner laminated glass panels by air gaps. An internal heated-air defrosting/defogging system, a deicing system, and an external liquid rain-removal system ensured good visibility in all weather conditions. The pilot's station included conventional aircraft controls and instruments.[15]

The crew cabin pressure could be set to 10,000- or 26,000-foot equivalent altitude pressurization. At altitudes below the pressure altitude selection, the cabin was essentially unpressurized. In theory, the pressurized cockpit allowed the pilot to operate in a standard flight suit with oxygen mask at altitudes below 50,000 feet, but a full pressure suit was normally worn to ensure crew safety under normal as well as emergency conditions. In an emergency the pilot could jettison the canopy and egress using a rocket-propelled ejection seat. The seat was usable at any altitude and Mach number within the flight envelope and included a drogue chute to stabilize the seat during descent. Man-seat separation occurred automatically on descent to, or at, barometric altitudes below 15,000 feet.[16]

The electronic compartment (E-bay) was located just aft of the pilot's station. This pressurized and air-conditioned space contained most of the communication and navigation equipment as well as the stability augmentation system, autopilot, flight reference, Mach trim, and other electronic systems.

The mission equipment bay (Q-bay) was located immediately aft of the E-bay and could be pressurized or unpressurized depending on specific equipment needs. This compartment provided space for installation of cameras and sensors, test packages, and/or ballast as dictated by mission requirements.

Air-conditioning equipment was located in the AC-bay, just aft of the Q-bay. This compartment housed most of the environmental control system equipment and the inertial navigation system. It also provided access to various circuit breakers and miscellaneous electrical components.

An in-flight refueling receptacle was located on top of the fuselage, just aft of the AC-bay. When de-energized, the receptacle doors formed the upper fuselage contour. When electrically actuated, the doors opened to reveal a trough to accept the aerial tanker's refueling probe.

Another set of doors on the upper side of the aft fuselage provided a cover for the drag chute compartment. The drag chute, along with the wheel-braking system, aided airplane deceleration during normal landings or aborted takeoffs.

The underside of the fuselage featured nose and main landing-gear wheel wells with hydraulically and mechanically actuated flush doors. The main gear wells also included insulated buckets to protect the tires from overheating while retracted during cruise.

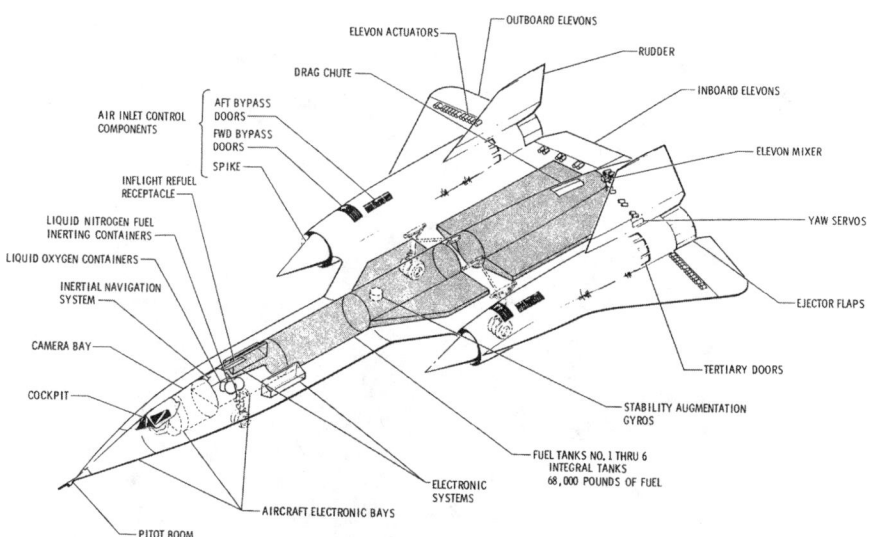

Every part of the A-12 was designed for maximum functionality. There was no excess space or unnecessary structural weight. (Courtesy of Lockheed Martin)

Remaining internal spaces in the fuselage were occupied by six integral fuel tanks. These, along with the wing tanks, provided a fuel capacity of 69,800 pounds. The fuel was pressure fed by two or more boost pumps in each tank. A cross-feed transfer system allowed the tank pumps to supply fuel to either engine. The airplane's tail-cone assembly contained a mixer for elevon control and a fuel dump port for use in the event of an in-flight emergency.

The wing was a thin, modified double delta with rounded tips. It was fully cantilevered, highly tapered and, in addition to the basic structure, incorporated inboard and outboard elevons to provide the combined aerodynamic functions of ailerons and elevators. Except where interrupted by the main gear well, the wing acted as an integral fuel cell between the leading edge and elevon support beams and spanwise between the fuselage and engine nacelle. The external surfaces of the upper and lower wing panels were beaded and corrugated to permit the skin and structure to expand and contract in response to temperature changes during flight.

At midspan each wing supported an engine nacelle containing rings and carry-through structure to support the outer wing. Both inner and outer wings were of multispar construction with chordwise stiffened skin panels attached to spanwise beams. The outer nacelle half (with attached outer wing) was hinged to open for engine access and servicing.

The tail group consisted of two inwardly canted, all-moving rudders and the inboard and outboard elevons. Each rudder was mounted on a fixed stub fin atop the rear portion of the engine nacelle. Contained within each stub fin, an electrohydraulic actuator moved the rudder through 20 degrees of travel

Several subscale models of the early A-12 design were tested on an outdoor RCS measurement range. (Courtesy of Lockheed Martin)

Eventually, RCS measurements of a full-scale pole model helped finalize the A-12 configuration. (Courtesy of Lockheed Martin)

on each side of the neutral position. The hydraulically actuated elevons were secured to the aft wing beam.[17]

AIRCRAFT SYSTEMS

In addition to mission-specific camera/sensor payloads, the Blackbird's designers had to accommodate all of the basic systems necessary to operate the airplane. These systems had to fit within the available space, which was dictated by aerodynamics and structural requirements, and also be capable of surviving the extreme environment created by high-speed, high-altitude flight.

Four independent systems (designated A, B, L, and R) supplied hydraulic power to the airplane to operate control surface actuators, landing gear, and other equipment. The A and B systems operated in parallel to supply hydraulic pressure to the flight controls, specifically to the seven actuating cylinders on each outboard elevon and three on each inboard elevon, as well as to two cylinders for each rudder. A dual servo unit, one for each movable flight control surface, controlled system pressure and return of fluid to the actuating cylinders.

The L and R systems supplied hydraulic power to the left and right inlet spikes and the forward and aft bypass doors on each nacelle. The L system also serviced the normal brake system, landing gear, main-gear inboard doors, nosewheel steering system, refueling door, and probe latches. The R system supplied hydraulic power to the alternate braking system, alternate nosewheel steering system, landing-gear-retraction system, and backup system for closing the main-gear inboard doors. Each system was served by its own hydraulic reservoir and fixed-angle, variable-volume piston pump. The left engine drove the A and L system pumps, while the B and R pumps were driven by the right engine.

The airplane's tricycle landing gear featured a dual-wheel configuration for the nose gear and triple-wheel configuration for the main gear. An actuating cylinder, located aft of the nose-gear strut, mechanically opened and closed the nose-gear doors, and controlled extension and retraction of the nose wheel. A hydraulic steering damper allowed the nose wheel 45 degrees of travel on either side of neutral. The pilot used rudder pedals in the cockpit to control the direction and degree of turn.

The main landing-gear outboard doors, linked to each gear strut, moved with the gear during extension and retraction. Hydraulic cylinders located inboard of the gear struts operated the main-gear assemblies and inboard gear doors. Each wheel assembly incorporated hydraulically actuated multiple disc brakes for landing deceleration.

The propulsion system consisted of two afterburning turbojet engines, one in each nacelle, and an axisymmetric inlet and air bypass system. Following early tests powered by two Pratt and Whitney J75 engines, the

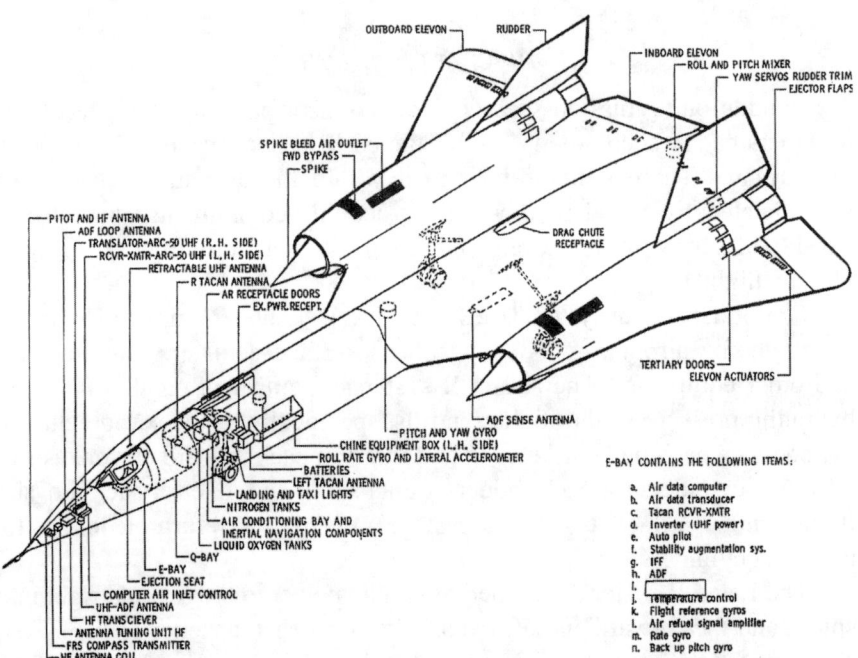

The final configuration of the A-12 was a single-place, twin-engine airplane with a modified double-delta wing and twin, inwardly canted tail fins. (Courtesy of CIA)

The A-12 had hydraulically actuated tricycle landing gear. When retracted, the main gear was stowed in insulated compartments to protect it from heating during high-speed cruise. (Courtesy of Lockheed Martin)

airplane was equipped with the J58, an axial-flow, gas-turbine powerplant with a nine-stage single-shaft compressor, can-annular combustion chamber, two-stage reaction turbine, and afterburner. The nacelle assembly included an exhaust ejector designed to operate in conjunction with a variable-area exhaust nozzle attached to the engine afterburner. Each air intake served as an external-internal compression inlet incorporating a movable spike to position the oblique shock wave and vary the inlet throat area. Movable bypass doors regulated inlet airflow to the engines and optimized inlet duct performance. An automatic air inlet control system included Mach, pitch and yaw sensors, a pitot-static probe, alpha-beta probe, angle and pressure-ratio transducers, inlet control computer, and hydraulic servos to actuate the spike and bypass doors. These elements were supplemented by boundary-layer bleeds and air bypass ducts to the exhaust ejector.

The fuel system was composed of controls, fuel tanks, plumbing and manifolds required to supply the engines with adequate fuel for takeoff, performance of mission requirements, and landing. Automatic fuel sequencing maintained the center of gravity (c.g.) within limits at all times. To maintain optimum c.g. during afterburner operation, the pilot controlled forward fuel transfer while the automatic system controlled aft fuel transfer. In an emergency, the pilot could dump fuel via the tail-cone vent. The system included emergency shutoff valves for both engines. A liquid-nitrogen inerting system and multiple relief valves maintained fuel-tank pressure.

An auxiliary circulating system employed the fuel as a heat sink for cooling the hydraulic and air conditioning systems and the engine accessory drive.

The fuel absorbed excess heat as it circulated through fuel-to-oil and fuel-to-air heat exchangers. Depending on its temperature, the fuel was directed into the engine feed lines or diverted back into the tank for cooling.

Two identical air-cycle refrigeration units conditioned bleed air from the engines for cooling and pressurizing the cockpit and equipment bays. Air-to-air and air-to-fuel heat exchangers provided primary cooling in the main landing-gear wells and engine nacelles. Air-to-fuel heat exchangers and air-cycle machines located in the AC bay provided secondary cooling. Both the cockpit and E-bay air conditioning systems were equipped with manual controls.

The cockpit pressurization schedule provided an unpressurized environment up to 26,000 feet, which was then maintained isobaric at that level for all higher altitudes. Both the E-bay and Q-bay were similarly maintained, but at 28,000 feet pressure altitude. For certain equipment, the Q-bay remained unpressurized throughout the mission. Pressurization was maintained by restricting air outflow, and provisions for cockpit pressure dump were available if necessary. The system also provided air for canopy and hatch sealing. The primary cooling system provided air at approximately 200°F for windshield defogging and defrosting. Air for the pilot's pressure suit was fully conditioned to maintain a comfortable temperature and had priority over cockpit cabin air. The suit pressurized automatically in the event of cockpit depressurization.

A full-power irreversible flight control system provided control-surface deflection in all three axes using hydraulic systems A and B. Either system could provide full deflection alone, but typically operated together. Special springs artificially provided pilot "feel" in each axis.

During a test flight in July 1962, the first A-12 dumps fuel from a vent in the tailcone. (Courtesy of Lockheed Martin via Jim Goodall)

Cockpit controls consisted of conventional stick, rudder pedals, trim control switches, and associated instrument displays, system indicators, annunciator lights, and circuit breakers. Dual cables transmitted pitch and roll inputs to the mixer assembly, a complex device used to combine the inputs into the desired elevon output. Each rudder was equipped with a closed-loop cable system to control yaw. Position of the main control surfaces was dependent on all flight control inputs including stick, rudder pedals, manual and automatic trim, stability augmentation system (SAS), and autopilot.

The automatic flight control system (AFCS) included a redundant three-axis SAS, two-axis autopilot, air data computer, and a Mach trim subsystem. Other associated equipment included an inertial navigation system (INS), flight reference system (FRS), hydraulic servos, and pitch trim actuator.

The SAS, an essential part of the basic control system, combined electronic and hydraulic equipment to augment the airplane's natural dynamic and static stability. It was optimized for cruise conditions but provided improved stability during all other flight conditions. The SAS constantly monitored aircraft attitude and provided control signals relative to the rate of change in all three axes, thus damping excessive changes in attitude. Because SAS corrections were applied through a series of servos, they were not apparent to the pilot at the control stick and rudder pedals.

A two-channel (pitch and roll) autopilot processed INS/FRS inputs and then applied the data through the SAS electronics to transfer valves for control-surface positioning. The pitch autopilot provided attitude hold or

The cockpit of the prototype A-12 as it appeared in January 1962, three months before first flight. Over the next six years, the control panel configuration was changed several times in response to suggestions by the pilots. (Courtesy of Lockheed Martin)

Mach hold. The roll autopilot provided attitude hold, heading hold, or automatic navigation (AUTO NAV). All autopilot and stability-augmentation-system inputs were summed (combined) prior to application to the respective transfer valves.

The Mach trim system compensated for the airplane's nose-down and nose-up tendencies while accelerating through Mach 0.2 to 1.5 by applying pitch trim to restore conventional stick forces and trim requirements at those speeds. The Mach trim system was operable any time the pitch autopilot was disengaged.

The electrical system consisted of two 30-KVA alternating-current (ac) generators mounted on gear boxes, one on each side of the airplane, and each was driven by its respective engine. A transfer system furnished variable-frequency, three-phase power to two ac buses. In the event of a single generator failure, the total ac load was transferred to the remaining generator.

Two 200-ampere transformer-rectifier (TR) units, obtaining power from each ac bus, provided direct current (dc) power. Two silver-zinc batteries provided emergency dc power in the event of a dual failure of both ac generators or both TR units.

A rocket-powered ejection seat served as an emergency escape system. The seat accommodated a fully pressure-suited pilot and incorporated life support and survival equipment including normal and emergency oxygen supply, backpack parachute, and seatpack survival kit. Seat restraints and safety features included an automatic opening lap belt, shoulder harness with inertia reel, knee guards and foot retention cables, and seat/man separation

Technicians prepare Article 121 for a test flight. Note F-104 safety chase plane. (Courtesy of Lockheed Martin)

device. The escape system was designed to provide safe egress at 65 knots ground speed on the runway up to the airplane's maximum speed and altitude conditions.[18]

Blackbird crews found it necessary to eject on 17 occasions in conditions ranging from takeoff to high-altitude cruise. Only four crewmen suffered fatal injuries. In one instance a pilot departed his SR-71A at an altitude of 80,000 feet while traveling at Mach 3, but without activating his ejection seat. The airplane simply broke apart around him. Fortunately his pressure suit and parachute system functioned properly despite the fact that he failed to initiate the normal ejection sequence. After freefalling to 15,000 feet, his chute opened automatically and he landed safely.[19]

For pilot life support the airplane was equipped with a liquid-oxygen (LOX) system consisting of two dewars, each containing 10 liters of LOX. This system converted LOX to gaseous oxygen through a temperature and pressure regulation process.

Radio communications and navigation systems provided air-to-air and air-to-ground voice transmissions, distance information to other aircraft or fixed ground stations, directional bearing to any given station, and other navigational information. An interphone system allowed communications between the cockpit and maintenance crew during ground operations and between the cockpit and tanker crew during aerial refueling. In two-seat Blackbirds the system allowed communication between crew members. A Dictet device recorded the pilot's microphone transmissions.

On January 25, 1966, this SR-71A was lost when it broke apart in flight at Mach 3 speeds and an altitude of 80,000 feet. The pilot survived despite being separated from the plane without his ejection seat. (Courtesy of Lockheed Martin)

A high-frequency (hf) radio provided long-range, two-way communication and operated in conjunction with a Birdwatcher system that automatically monitored the condition of various airplane systems. An ultrahigh-frequency (uhf) system provided short-range, two-way communications and direction finding for navigation. A low-frequency (lf) system provided long-range automatic direction finding while a tactical air navigation (TACAN) system provided short-range bearing, course deviation, and distance information relative to TACAN ground stations.

A self-contained inertial navigation system provided the principal navigation references for the airplane, entirely independently of electromagnetic radiation or other external references. The system provided attitude, true heading, command course, ground speed, distance, and geographic position data for automatic or manual navigation between points on the flight plan. The pilot could update position information periodically to correct for gyro drift by taking terrain fixes with a periscope that provided an optical display of terrain along the flight path or sun fixes with an optical device to measure sun azimuth angles for determination of true heading.

The flight reference system provided magnetic heading information and served as an alternate navigation reference. A gyro magnetic compass provided both gyro slaved and free gyro heading information while a gyro platform provided pitch and roll information.

A set of integrated flight instruments consisting of an attitude indicator, bearing-distance-heading indicator, and related signal-switching equipment displayed navigation information to the pilot. The indicators operated in

In flight and unpainted, the A-12 looked like a vision from the future as it streaked through the sky during early test flights. (Courtesy of Lockheed Martin)

conjunction with the inertial navigation and flight reference systems to provide data to the pilot.

An identification, friend or foe (IFF), transponder identified the airplane and its position with respect to an interrogation station. The unit received and decoded interrogation signals and then transmitted a coded response.

A radar beacon provided identification of the airplane and its position relative to ground and airborne X-band radar stations. The beacon also fortified the airplane's radar return to the receiver station, enhancing its target to increase the effective radar range.

A flight recorder system continuously recorded certain flight conditions. Operating in conjunction with a servo amplifier, it recorded airspeed, altitude, vertical acceleration, and magnetic heading vs elapsed time on an aluminum foil tape. This type of flight recorder was used only on the A-12, YF-12A, and M-21. The later SR-71 featured a more sophisticated mission recorder system.[20]

The SR-71 was also equipped with an astroinertial navigation system (ANS). In its primary mode of operation, the ANS used a gyro-stabilized inertial platform to sense aircraft motion and limited navigational errors through the use of a start-tracking system. In this mode, the system had a circular error probability (precision radius) of about 0.5 nautical miles for periods up to 10 hours. The ANS principally served as an automatic navigation system allowing the aircraft to accurately fly a predetermined flight path from a flight plan loaded into the navigation computer prior to takeoff. The plan could, if necessary, be modified prior to or during flight by redefining a series of destination points along the flight path.[21]

OXCART—PROGENITOR OF THE BLACKBIRD FAMILY TREE

The A-12 spawned a series of advanced airplanes that culminated in the SR-71. All variants were known as Blackbirds, and all but a few reached the prototype stage. Only the first and last variants, however, matured into operational systems.

The first A-12, known as Article 121, was trucked to the test site and reassembled following its arrival on February 27, 1962. While preparing for the maiden flight, technicians discovered the fuel tank sealant did not adhere properly to the titanium interior surfaces of the tanks. New sealant solved the problem of major leaks, paving the way for static engine runs.

By April 25, low- and medium-speed taxi tests were complete, and Lockheed test pilot Louis W. Schalk prepared for the first high-speed taxi run. He intentionally left the stability augmentation system as he accelerated down the runway. The airplane quickly attained takeoff speed and lifted off the ground, flying for about a mile. While airborne, the A-12 rocked wildly

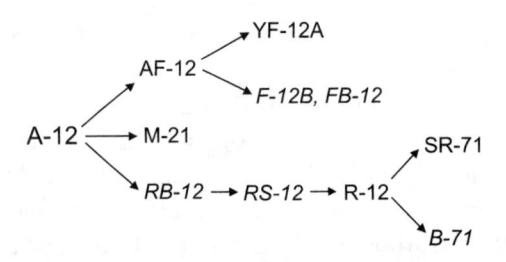

```
         ┌──→ YF-12A
      AF-12
   ┌──→    └──→ F-12B, FB-12
A-12 ──→ M-21              ┌──→ SR-71
   └──→ RB-12 → RS-12 → R-12
                           └──→ B-71
```

The A-12 gave rise to a complex "family tree" of Blackbird aircraft designs. Few of them, however, reached the production phase. (Courtesy of Lockheed Martin)

from side to side as Schalk wrestled it back to the ground on the emergency overrun beyond the departure end of the runway.

He later described his adventure:

> It all went like a dream until I lifted off. Immediately after liftoff I really didn't think I was going to be able to put the aircraft back on the ground

Article 121 made its official first flight in front of VIP guests on April 30, 1962. As can be seen here, it lacked the antiradar treatments that were applied to later production models. (Courtesy of Lockheed Martin)

safely because of lateral, directional and longitudinal oscillations. The aircraft was very difficult to handle but I finally caught up with everything that was happening, got control back enough to set it back down, and to chop engine power. Touchdown was on the lakebed instead of the runway, creating a tremendous cloud of dust into which I disappeared entirely. The tower controllers were calling me to find out what was happening and I was answering, but the UHF antenna was located on the underside and no one could hear me. Finally, when I slowed down and started my turn on the lakebed and re-emerged from the cloud of dust, everyone breathed a sigh of relief.

Despite the unexpected excitement, Schalk agreed to attempt the scheduled first flight two days later. He recommended, however, that the SAS dampers be turned on.[22]

The planned maiden flight on April 26, 1962, was generally successful but revealed some minor problems. As the airplane reached an altitude of about 300 feet, it began shedding fillet panels from the lower fuselage. This had little effect on flying qualities, and so Schalk completed a brief checkout of the aircraft. Per standard practice for a maiden flight, the landing gear remained down and locked throughout the flight. Schalk cycled the SAS dampers in all three axes without incident and completed his mission with an uneventful landing. On April 30, he flew a one-hour sortie in front of an audience of dignitaries, climbing to 30,000 feet and attaining a top speed of 392 mph. Just four days later, he expanded the airplane's flight envelope beyond Mach 1.0.[23]

Lockheed test pilot Louis Schalk prepares for a test flight in Article 121. Designer Kelly Johnson is barely visible at extreme right. (Courtesy of Lockheed Martin)

Over the next several months, the A-12 fleet grew as each airframe left the assembly line for the test site. The first five aircraft initially flew with J75 engines. Article 122 arrived in June but was not immediately used for flight testing. Instead, the airplane was raised, inverted, on a pylon for three months of radar-cross-section measurement tests. Article 123 arrived in August and flew in October. It was destined to become the first A-12 to crash when it was lost in a nonfatal accident on May 24, 1963. In November the A-12 trainer, Article 124, arrived at the test site. It had a second cockpit, for an instructor pilot, behind and above the student cockpit. Lockheed designers designated the trainer as the A-12T, but the pilots' official flight records (AF Form 5) list the airplane's mission design series designator as TA-12, a more conventional nomenclature for contemporary training aircraft. This airplane retained J75 engines throughout its service life and never attained Mach 3 speeds. Article 125 arrived in December 1962.

While awaiting the arrival of J58 engines, the A-12 fleet continued to operate with the J75 and fixed inlet spikes. This created some difficulties because the engine and inlet were mismatched, resulting in duct rumble (inlet airflow vibrations) as the A-12 approached Mach 2.0. A single J58 was eventually installed in Article 121, whereas the other nacelle held a J75. After building designers' confidence in the new powerplant, the airplane made its first sortie powered by two J58s in January 1963. Eventually the entire fleet, with the exception of Article 123 and Article 124, was retrofitted. Over the ensuing year eight more A-12 airframes joined the fleet.

The A-12T was the sole A-12 trainer model. It had a second cockpit, for an instructor pilot, mounted above and behind the student cockpit. (Courtesy of Lockheed Martin via Jim Goodall)

FORM FOLLOWS FUNCTION

The Blackbird fleet in April 1964 consisted of seven A-12 aircraft, the A-12T, and two YF-12A models. (Courtesy of Lockheed Martin)

In 1962 the first few A-12 airframes were flown unpainted, in bare metal finish. The following year black paint was applied to the nose, chines, inlet spikes, and leading and trailing edges of the wings to more effectively radiate heat. In 1964 the airplanes, with the exception of the A-12T, were painted entirely black. All subsequent variants adopted this paint scheme, thus earning the name "Blackbirds."

Over the next several years Lockheed and Central Intelligence Agency pilots conducted developmental and operational test and evaluation flights. The A-12 was cleared throughout its planned flight envelope and declared

In 1964, the A-12 aircraft were painted entirely black to more effectively radiate heat. At this time the airplane came to be known as the Blackbird. (Courtesy of Lockheed Martin)

mission ready. A sortie on November 20, 1965, included a maximum endurance flight of six hours and 20 minutes, including Mach 3.2 cruise at altitudes approaching 90,000 feet. Test and training flights continued for the purposes of maintaining pilot proficiency, developing mission equipment, and refining aircraft performance. Lockheed test pilot William Park performed a long-range sortie in December 1966, flying nonstop for 10,200 miles in just over six hours.

In May 1967 three A-12 aircraft were deployed to Kadena Airbase, Okinawa, Japan, for Operation Black Shield. Over the next 12 months the A-12 was flown on 27 operational reconnaissance missions over Southeast Asia and North Korea.

By this time the Air Force had a two-seat variant of the Blackbird. During the 1960s, when it became apparent that two fleets of high-performance aircraft would be conducting essentially identical missions for two agencies, the federal Bureau of the Budget began looking at ways to reduce operating costs. Ultimately President Lyndon B. Johnson accepted a recommendation to retire the A-12 aircraft following Operation Black Shield. By the end of the OXCART program, there were seven remaining A-12 aircraft and the single A-12T. By June 21, 1968, all had been placed in storage at Lockheed's Site 2 facility at Air Force Plant 42 in Palmdale, California.[24]

KEDLOCK—ADVANCED INTERCEPTOR

On March 16, 1960, Kelly Johnson traveled to Washington, D.C., to speak with Air Force officials about building an air-defense interceptor

Article 130, seen here in a 1963 photo, was used for test and training flights for the purposes of maintaining pilot proficiency, developing mission equipment, and refining aircraft performance parameters. (Courtesy of Lockheed Martin)

variant of the A-12. Called the AF-12, it was to be capable of launching air-to-air missiles at targets of varying altitude. Johnson's Air Force contacts provided him with information on the Hughes AN/ASG-18 search and tracking radar and the Hughes GAR-9/AIM-47 missile, asking Johnson if he could make use of this equipment in the AF-12. If he could, they were willing to propose the AF-12 as a standby air-defense fighter. Although there would not be an immediate production order, the Air Force officials were interested in starting a development program nonetheless. Johnson promised he could begin production of a prototype within a couple of years despite his OXCART commitments. By July 19, he had written a proposal for implementing a low-risk, minimum-cost approach to building the AF-12.[25]

Three months later Johnson received a letter of intent from the Air Force authorizing $1 million to "proceed with Plan 3A" (the official program name was KEDLOCK). As a result, Johnson planned to begin construction of the first AF-12, Article 1001, following completion of the sixth A-12. It would be built by a separate working group within the Skunk Works, under the supervision of project engineer Russ Daniel.

Johnson conceived the AF-12 as a modified A-12 airframe incorporating a fire control system coupled with the AN/ASG-18, the first U.S. coherent-pulse Doppler radar for long-range, look-down/look-up, and single-target attack. It employed a high average power, liquid-cooled, traveling-wave-tube (TWT) transmitter chain consisting of two TWT amplifiers in tandem to provide the desired gain and analog circuitry for generation and processing of the coherent high-pulse-repetition frequency wave form.

By June 21, 1968, all surviving A-12 airframes had been placed in long-term storage at Lockheed's facility in Palmdale, California. (Courtesy of Lockheed Martin)

YF-12A under construction at Burbank. Production of the Air Force interceptor prototype was segregated from the A-12 production line for security reasons. (Courtesy of Lockheed Martin)

The AF-12 design necessitated numerous changes to external configuration. A second crew position was added behind the cockpit to accommodate a fire control officer (FCO). Two infrared (IR) sensors, an integral part of the target tracking system, were placed on either side of the nose. The nose assembly itself was originally to be chined like that of the A-12, but was soon replaced by a radome with a circular cross section.

A mock-up review took place at Lockheed's Skunk Works facility in Burbank on May 31, 1961, causing Johnson to worry about program security.

"I was very concerned," he wrote in his log, "when I learned that some 31 people were coming, but the mock-up group from the Air Force consisted of about 15 people."[26]

The Air Force delegation told Johnson they were very pleased with the status of the mock-up and the information provided by Johnson's engineering group. Lockheed was ready to proceed with wind-tunnel model testing to verify the aerodynamics of the configuration.

The nose configuration, designed to accommodate the radar, altered the Blackbird's aerodynamics significantly and resulted in directional stability problems. Engineers resolved the problem by adding two small ventral fins to the engine nacelles and a large, hydraulically powered folding ventral fin on the centerline of the aft fuselage. Because of its size, the fuselage fin had to be folded to one side prior to takeoff and landing.[26]

By March 1962 officials from the CIA and even the White House were pressuring Lockheed to get the A-12 flying and operational as soon as possible.

FORM FOLLOWS FUNCTION 39

From above, the distinctive shape of the YF-12A nose assembly is evident. The chines were cut off abruptly just before the radome. (Courtesy of Lockheed Martin)

A forward fuselage mock-up of the production F-12B displays the final planned configuration with chines added to the radome and infrared search and tracking sensors (small white dome) below the cockpit. Courtesy of Lockheed Martin)

They asked that Johnson move the AF-12 prototypes to the end of the planned 10-airplane production run and assured him they would arrange for approval of this change from the Air Force.[27]

As A-12 airframes left the assembly line and entered flight testing, Johnson's team began to concentrate on completing the AF-12. In 1962, the Department of Defense instituted a common designation system for military aircraft, under which the AF-12 became the YF-12A. The first airframe was completed in mid-1963 and moved to the test site for final assembly.

Technicians spent several weeks installing the engines and testing various systems. They also conducted engine runs and hydraulic leak checks. Taxi tests revealed small problems with the brake and damper systems, but these were soon fixed. On August 7, 1963, Lockheed test pilot James D. Eastham completed the maiden flight in Article 1001. He found the YF-12A performed as well as the A-12 and had similar handling qualities. Two additional YF-12A airframes were soon added to the test program.

President Lyndon Johnson publicly announced the existence of the YF-12A on February 29, 1964. At Kelly Johnson's request, however, he called the airplane A-11 because that designation denoted the non-anti-radar design. This needless bit of subterfuge has plagued historians for years as various authors have referred to the airplane using the essentially fictitious designation.

Within minutes of the president's announcement, the first two YF-12A airframes were on their way to Edwards Air Force Base. The third, Article 1003, completed its maiden flight 12 days later.[28]

The first YF-12A landing at the end of its maiden flight on August 7, 1963. All Blackbirds were equipped with a drag chute to assist with deceleration. (Courtesy of Lockheed Martin)

Air Force crews soon began testing the aircraft's weapon systems while attempting to solve troublesome problems with transonic acceleration and various subsystems. The following year several official speed and altitude records were set in the aircraft. Having the YF-12A in the public eye provided cover for the CIA's A-12 operations. If a civilian happened to spot the A-12 in flight, they would simply assume it to be a YF-12A.

The first YF-12A was badly damaged on August 14, 1966 in a landing accident. Several years later the rear half of the airplane was salvaged to build the SR-71C, a replacement for an SR-71B trainer that was lost in a crash. The two remaining airframes continued to serve as prototypes for a planned operational version of the interceptor, designated F-12B, but by this juncture the program was in serious trouble.[29]

Secretary of Defense Robert McNamara was engaged in a bitter feud with the Air Force over appropriation of defense funds, and he specifically targeted planned production of the F-12B in his cost-cutting measures. On several occasions he denied the Air Force access to $90 million that already had been appropriated by Congress for F-12B production. The YF-12A operation at Edwards was curtailed, and, by August 1966, Lockheed had laid off half of its test personnel.

In March 1967, Colonel Benjamin Bellis, director of the F-12/SR-71 System Program Office (SPO), led Kelly Johnson to believe there was still a chance the Air Force would go through with procurement of the F-12B, and eight months later Johnson was asked about the feasibility of converting the stored A-12 airframes and 10 of the SR-71 aircraft into interceptors. In December, however, the SPO directed Johnson to terminate all F-12-related studies.

The YF-12A was designed to carry Hughes AIM-47 air-to-air missiles in weapon bays located in the forward fuselage chines. (Courtesy of Lockheed Martin)

Despite cancellation of F-12B production, the YF-12A saw continued use as a test aircraft in a joint program involving the Air Force and NASA. (Courtesy of Lockheed Martin)

On January 5, 1968, Air Force officials informed the Skunk Works that the F-12B program was cancelled. The YF-12A test program ended on February 1, and the Air Force directed Lockheed to destroy all A-12/F-12/SR-71 tooling to prevent future production of the aircraft. The surviving YF-12A aircraft soon joined the A-12 airframes in storage at Palmdale, but not for long.[30]

On June 5, 1969, NASA and Air Force officials signed an agreement for a joint research program using Articles 1002 and 1003. Air Force test researchers planned to conduct tests involving the tactical performance and support requirements of an advanced interceptor while NASA researchers wanted to explore propulsion systems, aerothermoelasticity, and flight-dynamics characteristics of supersonic cruise aircraft.

The new program spanned a decade and produced a wealth of data with applications to future aircraft designs. Article 1003 was lost in a nonfatal accident in June 1971, leaving only one remaining YF-12A. Article 1002 made its final flight on November 7, 1979, when it was ferried to Wright–Patterson Air Force Base, Ohio, for permanent display in the U.S. Air Force Museum.[31]

TAGBOARD — MACH 3 DRONE

In early 1962 Kelly Johnson met with Eugene Fubini of the Department of Defense Office of Research and Engineering to discuss the possibility of designing a drone aircraft capable of operating in the same speed and altitude range as the A-12. Such a vehicle would give the president a choice between manned or unmanned reconnaissance platforms to overfly sensitive areas. Johnson asserted that such a development was entirely feasible and suggested the drone could be launched from a modified version of the A-12.

Joseph Charyk, director of the National Reconnaissance Office, saw no inherent political value in using a drone and noted that the vehicle's capabilities would be no better than those of the A-12. He recognized, however, that such a scaled-down vehicle would be significantly harder for enemy radar to detect. By October 5, Robert McNamara and Director of Central Intelligence Allen Dulles agreed to proceed with development of the drone under Project TAGBOARD. Management responsibility for the program was assigned to the CIA with support from the Air Force.

Several days later, Kelly Johnson submitted a written proposal for a feasibility study regarding the Q-12, as he called the drone. It was accepted, and Johnson drew up plans for a ramjet-powered vehicle that would be launched for a dorsal pylon atop the fuselage of a modified A-12-type aircraft. The major modification to the OXCART design included provisions for a launch control officer (LCO) seated in a compartment behind the cockpit. The drone rested on a pylon mounted on the aircraft's centerline between the twin vertical stabilizers.

On February 28, 1963, Johnson received approval to produce 20 drones (Article 501 through Article 520) and two launch aircraft (Article 134 and Article 135). As construction began, Johnson saw the two vehicles as "mother" and "daughter." To avoid confusion with the single-seat reconnaissance jet, he designated the mothership M-21 (with "M" for "mother" and

The D-21 drone was pylon-mounted atop the M-21 launch aircraft. For captive tests, an aerodynamic tail cone was installed on the drone. (Courtesy of Lockheed Martin)

reversing the numerals). The drone became the D-21, and the mated combination was called MD-21.[32]

The M-21 had generally the same performance and handling characteristics as the A-12. To stay within the aircraft's capabilities while maintaining optimal attachment loads on the D-21, maneuvering loads were limited to no more than 2.0 grams for the mated configuration. Most rolling maneuvers were prohibited for the MD-21 to prevent adverse side loads on the D-21. The only aileron maneuver allowed for the mated configuration consisted of a steady turn, performed within the airplane's capabilities.[33]

The D-21 was almost 43 feet long with a nearly 20-foot wing span and about seven feet tall. As a result of shape and materials, it had a small radar cross section. Maximum gross takeoff weight was 11,000 pounds. It had a design cruise speed of Mach 3.35 at 80,000 to 90,000 feet and a range of 1439 miles.[34]

The drone featured a semimonocoque design using rib and skin construction. It was built primarily of Beta-120/Ti-13V-11Cr-3A1 titanium, although the tip of the inlet cone was made of a nickel alloy. To reduce RCS, many components were manufactured from the same sort of plastic laminates used on the A-12. These included wing leading-edge panels, elevons, elevon fences, telemetry and command antenna fairings (silicone–asbestos), camera hatch liner and parachute compartment (epoxy–fiberglass), and various antenna windows (silicone–fiberglass).

The interior of the fuselage was divided into four sections. The first included a circular air inlet with a fixed spike and air bypass system. Next, the equipment bay contained a hatch housing the camera and its parachute recovery system, which could be jettisoned, as well as navigational equipment and

The drone's fuselage was built as a semimonocoque design using rib and skin construction with titanium as the primary structural material. (Courtesy of Lockheed Martin)

guidance electronics. Nonrecoverable items housed in this section included auxiliary power and air-conditioning systems, destruct system, battery, relays, wiring, and components of the fuel, hydraulic and liquid nitrogen systems with interconnecting plumbing. The third section contained three integral fuel tanks, interconnected to provide fuel to the engine efficiently. The fourth section housed the engine itself with its integral tailpipe and exhaust nozzle. The engine inlet air duct extended from the nose section aft through the fuselage to the engine section.

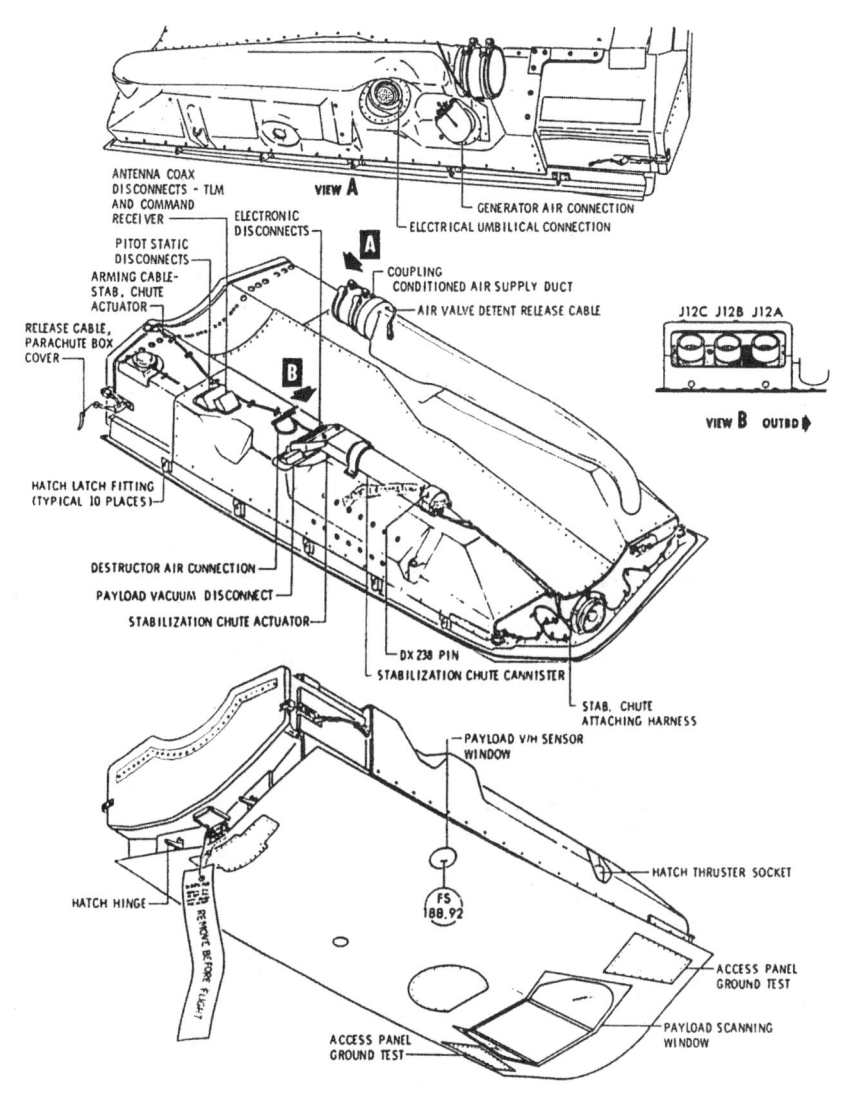

The D-21 was equipped with a recoverable camera hatch that was jettisoned following completion of its mission. (Courtesy of Lockheed Martin)

46   PETER W. MERLIN

The drone's primary wing structure consisted of sheet titanium skins covering titanium ribs developed from the fuselage frame sections. The wing leading-edge sections, chines, and elevons consisted of integrally bonded assemblies made up of a honeycomb core, molded edge, and silicone-asbestos laminated plastic upper and lower skins. These removable plastic sections were attached at their inboard edges to the wing beam and on their sides to tie-through ribs, spliced to the beam's front face.

Each forward chine section widened into a true wing as it extended aft to a removable torque box assembly, terminating in the elevon. The elevon was attached to the lower surface of the torque box beam by means of a continuous

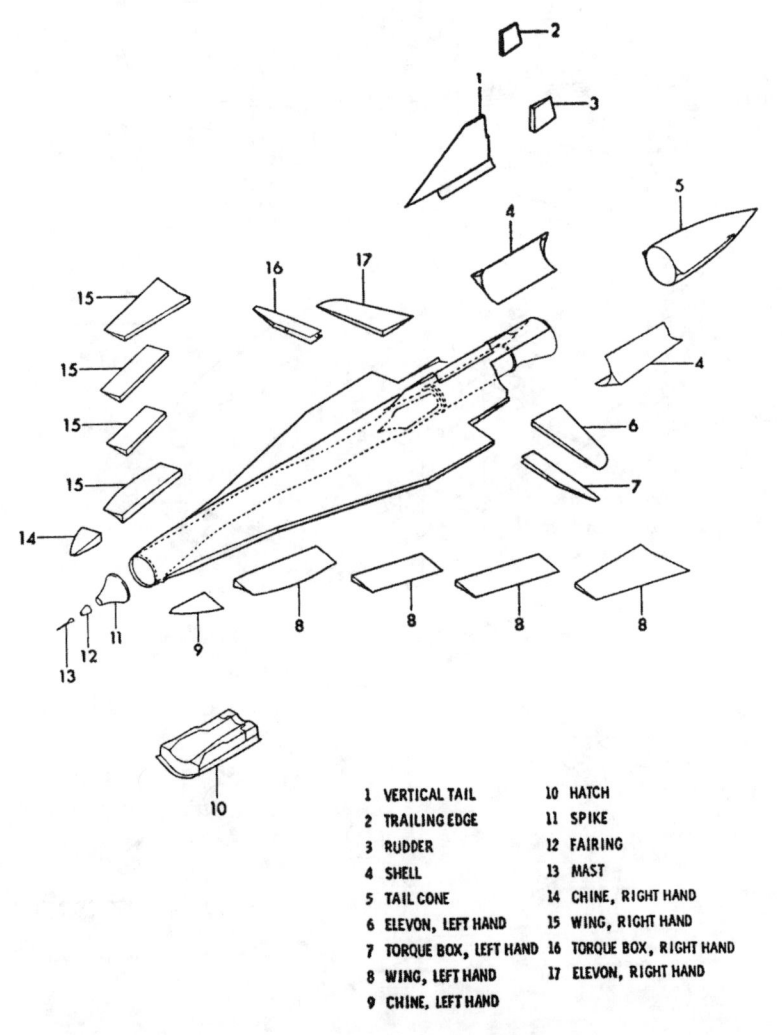

1	VERTICAL TAIL	10	HATCH
2	TRAILING EDGE	11	SPIKE
3	RUDDER	12	FAIRING
4	SHELL	13	MAST
5	TAIL CONE	14	CHINE, RIGHT HAND
6	ELEVON, LEFT HAND	15	WING, RIGHT HAND
7	TORQUE BOX, LEFT HAND	16	TORQUE BOX, RIGHT HAND
8	WING, LEFT HAND	17	ELEVON, RIGHT HAND
9	CHINE, LEFT HAND		

Major components of the D-21 are shown here, including an aerodynamic tail cone that was used only during captive test flights. (Courtesy of Lockheed Martin)

hinge and functioned as both elevator and aileron. The elevon structure included metal control fittings for attachment to the wing. Hydraulic actuators in the torque box beams worked through mechanical linkages to move the elevon surfaces up 17 degrees or down 18 degrees in coupled or independent motions.

The tail group consisted of a fixed vertical fin and hinged rudder, both of titanium rib and skin construction. The fin housed the fuel vent line, rudder torque tube, and hydraulic servo controls. The fixed fin was attached to a stub fin atop the aft section of the fuselage.[35]

The D-21 was powered by an internally mounted Marquardt XRJ43-MA20S-4 ramjet engine. With no inlet structure of its own, it relied on the drone's inlet system for air intake. The engine's flame holder was designed for stable combustion at extremely high altitudes and low atmospheric pressures. Fueled with PSJ-100 and ignited using triethyl borane (TEB), the engine was designed to operate continuously during missions lasting more than 1.5 hours. Until this time no ramjet had ever powered a vehicle for longer than 15 minutes.[36]

The engine was composed of several major assemblies including the nose cone, fuel control unit, forward inner body, main structure, aerodynamic grid, fuel manifold, burner, combustion chamber, and TEB ignition system.

Submerged within the drone's fuselage, the ramjet was supplied with high-pressure, low-velocity (subsonic) airflow from the supersonic inlet located in the vehicle's nose. The air was compressed in several steps through a process called aerodynamic diffusion. First, supersonic air was compressed and slowed by oblique shock waves as it entered the circular inlet around an external fixed spike. The decreasing cross-sectional area within the inlet cowl further compressed the air to subsonic speeds. The final compression phase occurred as the subsonic flow stream passed through ducting with incrementally increasing cross-sectional area. At the engine, the air flowed around the nose cone and innerbody before being evenly distributed by the aerodynamic grid nozzles. The air flowed past the manifold, mixed with fuel, then entered

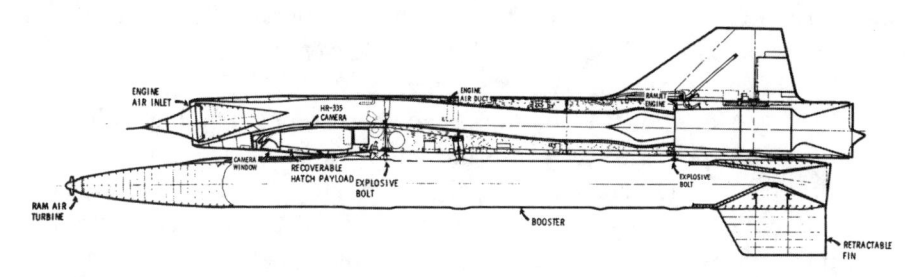

The D-21 was powered by a ramjet engine, submerged deep within its fuselage. The "B" model needed a DZ-1 rocket booster to propel it to Mach 3 speeds. (Courtesy of Lockheed Martin)

When the D-21 was mated to the M-21 launch aircraft, the combination was called the MD-21. During early captive flights, the drone was fitted with aerodynamic fairings on its nose and tail. (Courtesy of Lockheed Martin)

the burner gutters and ignited in the combustion chamber. The thrust force of the expanding gases accelerated the burned mixture to supersonic velocity as it ejected through the exit nozzle.

Although many components of the engine were made of steel, titanium, and aluminum alloys, the nose-cone assembly consisted of a semiconical magnesium-thorium (magthor) alloy fairing with a steel tip. The forward inner-body assembly, main structure, fuel inlet fitting, total pressure probe, aerodynamic grid, fuel control unit, and outer burner assembly were also made of magthor, a mildly radioactive alloy.[37]

Lockheed workers completed construction of a full-scale D-21 mock-up on December 7, 1962. Because it included an accurate representation of the drone's antiradar characteristics, the mock-up saw service as a pole-mounted model for RCS measurements. This resulted in modifications and design refinements to correct deficiencies. While these were undertaken, the mock-up was also used for fit checks of the camera system.

By early May 1963, Skunk Works engineers were struggling with aerodynamic loads problems. Because the wind-tunnel model of the MD-21 did not accurately represent the actual configuration, dynamic pressures in the wrong flowfields yielded confusing loads data. The configuration was finalized by October, and Johnson proposed that the load problems could be alleviated by letting the D-21 float at a zero-moment incidence when attached to the M-21, thereby concentrating the loads at one point and reducing them.

To alleviate load problems, the D-21 was mounted so that it floated at a zero-moment incidence when attached to the M-21. (Courtesy of Lockheed Martin)

On December 31, Johnson noted in his log that wind-tunnel data indicated there would be launch difficulties. "Going through the fuselage shock wave is very hard," he wrote. He believed it best to launch the drone at full power, but acknowledged there still were problems regarding the fuel-air ratio to the engine and engine flameout at that condition. Subsequent wind-tunnel tests highlighted the need for the M-21 pilot to initiate a pushover at launch because the pylon was too short. "This was done for structural reasons, but got us into aerodynamic troubles which weren't recognized at once," he wrote.[38]

By May 1964, the D-21 continued to be plagued by launch difficulties, transonic drag deficiencies, and equipment problems, but Johnson was confident these problems could be overcome. Construction of the first M-21 (Article 134) was well underway and a launch pylon had been developed to carry the 11,000-pound drone. The low-drag pylon, mounted on the mothership's dorsal centerline, contained a primary support post with main and secondary locking hooks, provisions for emergency pneumatic jettison, and plumbing to provide fuel for cooling the drone's subsystems and topping of its tanks prior to launch.

Fit checks of the mated MD-21 configuration took place in June, using Article 134 and the first production D-21 (Article 501). Final assembly of both vehicles was soon complete. On August 12, the M-21 was delivered to the test site for initial checkout work. The maiden flight of the MD-21 took place on 22 December, coincidentally the same day as the maiden flight of the SR-71.[39]

Following several successful launches, the second M-21 (seen here) was lost in a mid-air collision when the drone suffered an unstart immediately after release. (Courtesy of Lockheed Martin via Jim Goodall)

Initial drone separation trials began in March 1966. The flight profile required the M-21 pilot to attain a speed of Mach 3.12 at an altitude of 72,000 feet and commence a gentle pull-up before pushing the nose over to maintain a steady force of 0.9 g. With the drone's ramjet operating, the launch control officer initiated pneumatic separation of the drone. On its second free flight, the drone attained a speed of Mach 3.3 and altitude of 90,000 feet. Although it demonstrated the planned performance characteristics, the D-21 still suffered a few technical problems. Additionally, the launch maneuver was risky for the flight crew.

On April 29, 1966, in the wake of an order for a second batch of drones, Johnson made a formal proposal to Strategic Air Command Headquarters for a modified drone (eventually designated D-21B) that would be launched from beneath the wing of a B-52H bomber. This, he said, would provide greater safety, reduced costs, and expanded deployment range. To propel the D-21B to ramjet ignition speeds (around Mach 3.0), the drone would require a rocket booster for the initial flight phase following launch.

On its third launch from the M-21, the drone flew 1842 miles and executed eight programmed maneuvers. This spectacular success was marred only by its failure to eject the camera package as a result of an electronic malfunction.

On July 31, however, Johnson's worst fears were realized during the fourth launch attempt. After the drone lifted off the pylon, its ramjet unstarted. The D-21 rolled over on its side and plunged into the mothership's aft fuselage. The M-21 pitched up and broke apart, sending debris plummeting toward the Pacific Ocean. Although both crewmen ejected from the stricken craft, the launch control officer perished before rescue forces arrived.

Under Project SENIOR BOWL, two B-52H aircraft were modified to launch a variant of the drone called the D-21B. (Courtesy of Lockheed Martin via Jim Goodall)

As a result of the tragedy, Johnson cancelled further use of the MD-21. The remaining D-21 drones were modified to the D-21B configuration, and two B-5H aircraft were configured as launch platforms under Project SENIOR BOWL. The D-21B had a less sensitive inlet configuration to mitigate the unstart problem and was designed to hang from a pylon on the wing of the B-52. A 44-foot-long Lockheed DZ-1 solid-fueled rocket booster was attached beneath the drone. The booster weighed 13,286 pounds and produced 27,300 pounds of thrust during an 87-second burn. The D-21B experienced sudden acceleration to Mach 3.2 as its surface temperature rapidly went from −58°F at launch to 428°F at cruise, resulting in severe thermodynamic stress on the airframe.

To accelerate the D-21B to ramjet ignition speeds, a solid-fueled rocket booster was attached to the underside of the drone. A ram-air turbine on the booster's nose provided electrical power and hydraulic pressure to the drone prior to launch. (Courtesy of Lockheed Martin)

Between November 6, 1967, and March 20, 1971, the D-21B drones flew 12 test missions and four operational sorties. Each mission cost approximately $5.5 million, and the operational missions failed to produce the expected results because the camera packages were not recovered. On July 23, 1971, the program was cancelled, and all remaining drones placed in storage.[40]

SENIOR CROWN—STRATEGIC RECONNAISSANCE

On September 14, 1960, Johnson began work on a bomber version of the Blackbird that he called the RB-12. His proposal resulted from the reported development of small, high-yield nuclear warheads. He suggested four hypothetical 400-pound bombs, or a single large bomb, could be carried inside the airplane's fuselage without compromising fuel load. No aerodynamic changes were required, and the radar-attenuating features remained intact.

Johnson pitched his proposal to the Air Force, emphasizing the airplane's performance and survivability characteristics. He also had his Skunk Works team construct a mock-up of the RB-12 forward fuselage section. General Curtis LeMay and General Thomas Power visited Burbank in July 1961 and reviewed mock-ups of both the RB-12 and AF-12. Although Johnson found the Department of Defense more interested in the bomber than the interceptor, the RB-12 never went beyond the mock-up stage. It was ultimately rejected because it was seen as a threat to North American Aviation's XB-70, the proposed replacement for the B-52.

During September 1962, Johnson began exploring what he called a "common market" version of the A-12. A single airframe configuration, known as the R-12 Universal airplane, would serve as the basis for a reconnaissance, recon/strike, or interceptor variant, depending on customer needs. This, he

The SR-71, originally called the R-12, included many design elements in common with earlier Blackbird variants. (Courtesy of Lockheed Martin)

believed, would greatly simplify production. "It eliminates the necessity of the Air Force deciding which version they want to buy," he wrote in his log.[41]

In February 1963, Lockheed was given precontractual authority to build six R-12 airframes with the understanding that an additional order of 24 was forthcoming. The details of the contract allowed the Air Force and CIA to share the financial burden.

Parallel development of the A-12 and R-12 fueled a debate in the Pentagon as to the need for two similar reconnaissance platforms. The Air Force used the opportunity to press its case that it should have sole jurisdiction over such a mission. This eventually doomed the A-12 to cancellation.

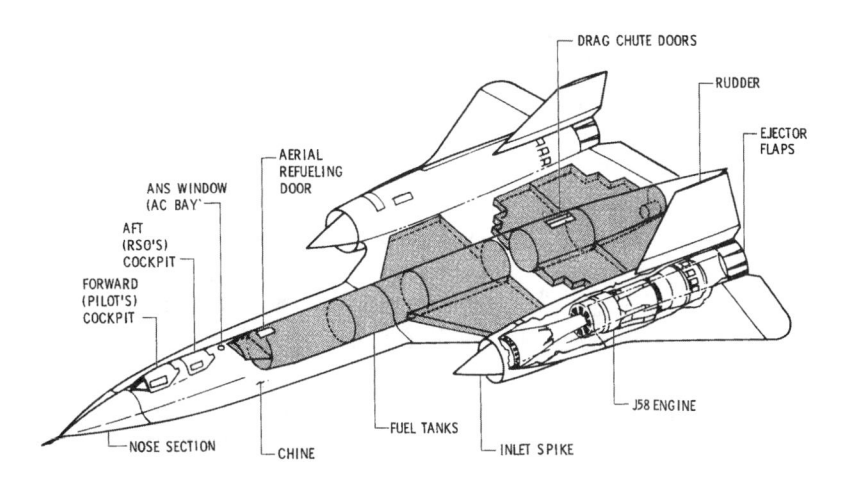

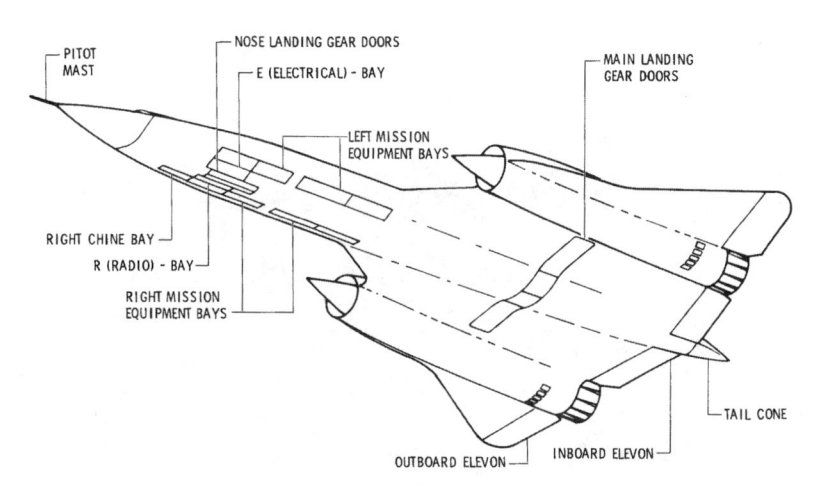

To the untrained eye, the SR-71 looked nearly identical to the A-12, but there were numerous significant differences. (Courtesy of AFFTC History Office)

As the Skunk Works pressed ahead with the R-12, the airplane's configuration diverged noticeably from that of the A-12. The most obvious difference was the second crew position, behind the cockpit, to accommodate the reconnaissance systems operator (RSO). The fuselage was lengthened slightly to make room for additional fuel capacity, and the tail cone was extended slightly. The nose chines were broadened to improve cruise characteristics and compensate for loss of directional stability as a result of the change in length. There were numerous internal changes as well with regard to various subsystems.

Air Force officials showed interest in the RS-12 reconnaissance/strike variant. Consequently, Johnson worked on systems and structural issues related to weapons carriage.

All of Lockheed's R-12 and RS-12 activities had been undertaken with great secrecy. On July 24, 1964, this changed when President Lyndon Johnson announced the development of "a major new strategic manned aircraft system" to provide Strategic Air Command with worldwide reconnaissance capabilities. He repeatedly referred to the new airplane as the SR-71, a designation that soon became official.

By the end of October, the first SR-71 airframe (Article 2001) had been moved to Lockheed's plant in Palmdale for final assembly. Engine runs were initiated in mid-December, quickly followed by taxi tests. Finally, on December 22, 1964, the SR-71 made its maiden flight, remaining airborne for more than an hour and achieving speeds in excess of 1000 miles per hour.

Although initial testing of the SR-71 was well underway, the Skunk Works had not yet abandoned the RS-12 concept. In April 1965, the demise

The first SR-71A (Article 2001) made its first flight on December 22, 1964. It was used exclusively for developmental testing. (Courtesy of AFFTC History office)

The SR-71B served as a trainer, with a raised instructor's cockpit. (Courtesy of Lockheed Martin)

of the XB-70 emboldened Kelly Johnson to present a small contingent of Air Force generals with a design for a bomber version of the SR-71 (known in-house as the B-71). It was destined, however, to remain a paper airplane.

SR-71 airframes began to roll off the assembly line in quantity, with the first few assigned to developmental testing. By late 1967, a total of 31 airframes had been delivered to the Air Force. Two of these were SR-71B trainer models (Articles 2007 and 2008) with a raised instructor's cockpit in place of the reconnaissance-systems-operator position. After Article 2008 crashed in January 1968, it was replaced with a trainer built from the aft fuselage of the first YF-12A (Article 1001) and the forward section of a structural test article (with added instructor's position). The new trainer was called the SR-71C (Article 2000).

During their service lives, 12 of the SR-71 airframes were lost to accidents. The operational fleet was retired twice, first in 1990. A few aircraft were reinstated into service, but retired again in 1997. NASA used several as research platforms and became the final agency to fly the Blackbird. The last SR-71 flight took place on October 9, 1999.

Most surviving airframes were placed on display in various museums around the United States and one in the United Kingdom. Article 2031, the last SR-71A built, was placed on permanent display at NASA Dryden Flight Research Center, Edwards, California, where it made its final flight.[42]

Chapter 3

NECESSITY IS THE MOTHER OF INVENTION: CONSTRUCTION AND MATERIALS

Lockheed engineers faced unique challenges in designing and building the Blackbirds. Aerodynamic friction and continuous engine operation during high-speed flight subjected some parts of the airplane to temperatures as high as 1050°F. Average surface temperatures ranged from 462 to 622°F. This precluded the use of aluminum as a basic structural material. The Skunk Works team turned instead to titanium, stainless steel, and other advanced alloys, as well as to high-temperature plastics.[43]

BUILDING THE BLACKBIRDS

Kelly Johnson received word on August 28, 1959, that the A-12 had been selected for production. The following day Richard Bissell of the Central

Lockheed technicians faced considerable challenges while constructing the A-12. Extensive use of exotic alloys and high-temperature plastics was unprecedented in aircraft production. (Courtesy of Lockheed Martin)

1959	1960	1961	1962	1963	1964	1965	1966	1967	1968
♦ LIMITED GO-AHEAD									
♦ FULL GO-AHEAD									
		♦ SCHEDULED 1st FLIGHT							
			♦ 1st FLIGHT WITH TWO J75 ENGINES						
			♦ 1st FLIGHT WITH ONE J58 ENGINE						
			♦ 1st FLIGHT WITH TWO J58 ENGINES						
				♦ MACH 3.0 ACHIEVED					
				♦ MACH 3.2 ACHIEVED					
					♦ LIMITED IOC DECLARED				
				1st OPERATIONAL MISSION ♦					
					LAST OPERATIONAL MISSION ♦				
					PROGRAM TERMINATION ♦				

A timeline for the OXCART program illustrates the milestones achieved over a period spanning nearly a decade. (Courtesy of Lockheed Martin)

Intelligence Agency gave him a limited go-ahead for preliminary work. Lockheed had four months to conduct wind-tunnel model tests and build a full-scale mock-up. On January 30, 1960, Johnson received full authorization to proceed with the design, manufacture, and testing of 12 airplanes.[44]

When Bissell authorized $4.5 million for the initial effort, he also made it clear that the CIA would conduct its business with Lockheed as it had during the U-2 program. In other words, Johnson would be given a fair amount of autonomy, enabling the Skunk Works team to control costs and meet rational program milestones and objectives. In return Bissell demanded that Lockheed exercise the greatest possible ingenuity to improve the airplane's antiradar characteristics, make no multimillion-dollar material commitments until complete funding was authorized, and maintain a higher level of security than that of the U-2 program.[45]

Once funding was fully authorized, Johnson's team began work in earnest. The first efforts involved selection of structural materials. Only titanium and steel could withstand the expected operational temperatures, but steel was extremely heavy, and Lockheed had little experience with lightweight stainless-steel honeycomb structures. Skunk Works technicians realized aged B-120 titanium had nearly twice the strength/density ratio of stainless steel per cubic inch, but weighed approximately half as much. They also found they could manufacture titanium structures with fewer parts using conventional construction methods than were necessary with steel. High-strength composites were not available in 1960. Although the Blackbirds featured an abundance of composite plastic laminates as antiradar treatments, those substances were not used as primary structure.

Initial material tests were not promising. When a representative titanium wing panel was heated to the expected in-flight temperatures using a thermal

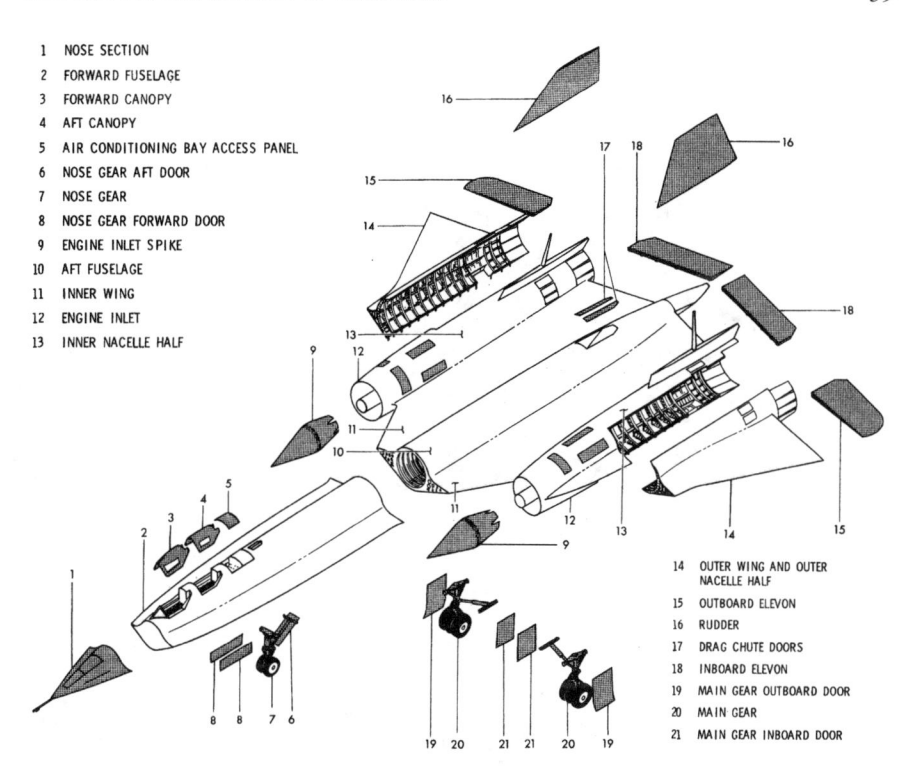

#	
1	NOSE SECTION
2	FORWARD FUSELAGE
3	FORWARD CANOPY
4	AFT CANOPY
5	AIR CONDITIONING BAY ACCESS PANEL
6	NOSE GEAR AFT DOOR
7	NOSE GEAR
8	NOSE GEAR FORWARD DOOR
9	ENGINE INLET SPIKE
10	AFT FUSELAGE
11	INNER WING
12	ENGINE INLET
13	INNER NACELLE HALF
14	OUTER WING AND OUTER NACELLE HALF
15	OUTBOARD ELEVON
16	RUDDER
17	DRAG CHUTE DOORS
18	INBOARD ELEVON
19	MAIN GEAR OUTBOARD DOOR
20	MAIN GEAR
21	MAIN GEAR INBOARD DOOR

Most of the major airframe components of the Blackbirds were assembled from titanium alloys. Composite plastic laminates were abundant in areas that required radar cross section reduction but were not used as primary structure, except in the inlet shock cones and tail fins. (Courtesy of Lockheed Martin)

test fixture, technicians watched in horror as the panel warped into an unacceptable shape. Undaunted, they manufactured a new panel with chordwise corrugations in the outer skins and reran the tests with satisfactory results. At design temperatures the corrugations merely deepened a few thousandths of an inch and then returned to their original shape as the metal cooled. The modification was incorporated into the design and became one of the Blackbird's signature features. Johnson later joked that he was accused of "trying to make a 1932 Ford Trimotor [an airplane surfaced with corrugated tin] go Mach 3."

Another structural test unit represented the airplane's forward fuselage and cockpit. It was used to determine the thermal effects on more than 6000 parts including the complex pilot's canopy. These tests also contributed to the development of cockpit cooling systems.[46]

While building the cockpit test article, the Skunk Works team discovered that the first batch of heat-treated titanium parts was extremely brittle. In fact, a part dropped on the shop floor actually shattered. Of the first 6000 pieces made from Beta B-120 titanium, 95% were rejected.[47]

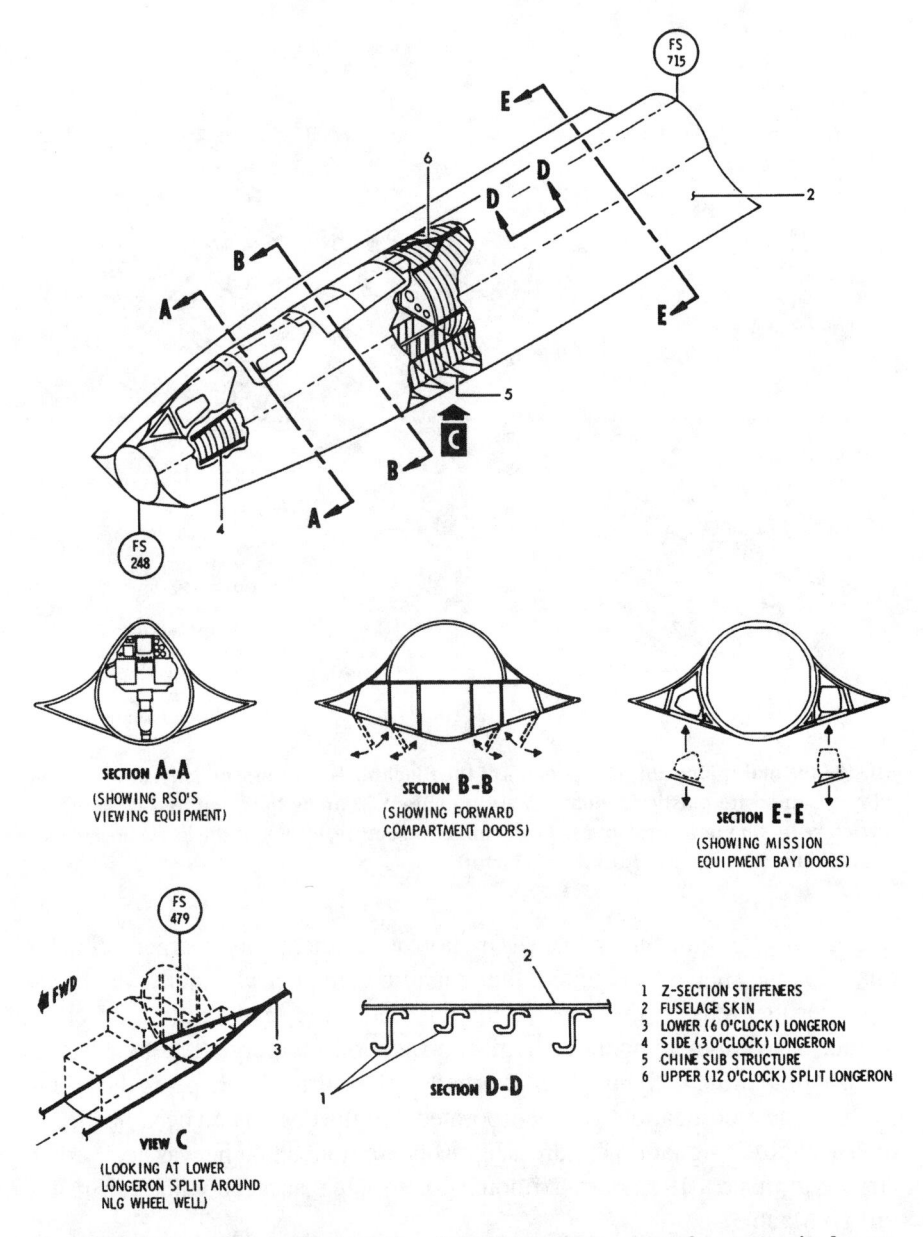

The forward fuselage consisted mostly of titanium alloys except for composite honeycomb skin panels supported by the chine structure. (Courtesy of U.S. Air Force)

At first it was thought the problem might have resulted from hydrogen embrittlement during the heat treatment process, but that theory could not be verified. Ultimately technicians resolved the issue by discarding the entire acid pickling setup and replacing it with one identical to that used by the supplier, Titanium Metals Corporation.

Next, Skunk Works staff developed a complex quality control program. For every batch of 10 or more parts, three test coupons (sample pieces) were subjected to identical heat treatment. One coupon was tensile tested to failure to provide stress-strain data. The second received a quarter-inch cut on one edge and then bent around a mandrel at the cut. If the coupon could not be bent 180 degrees at a radius of x times the sheet thickness (where x is a function of the alloy used and the stress-strain value of the piece) without breaking, it was rejected as too brittle. The third coupon served as a spare in case reprocessing was required. For large forgings such as landing-gear components, as many as 12 sample coupons were tested.[48]

Titanium rivets turned out to be among the most difficult items to manufacture. In the early days of the program, it was nearly impossible to obtain pure enough samples of the metal in wire form. For this reason titanium rivets, bolts, and other fasteners were initially very expensive. The price

The forward fuselage of the first SR-71 under construction. Note lightweight titanium chine structure. (Courtesy of Lockheed Martin)

gradually lessened as these items saw wider use throughout the aircraft manufacturing industry.[49]

Titanium alloys presented numerous challenges because of adverse chemical reactions between the metal and various materials and compounds commonly used in aircraft construction and maintenance. To prevent corrosion and stress-corrosion cracking at high temperatures, it was imperative that titanium parts be prevented from contacting such materials as cadmium, mercury, fluorine, chlorine, bromine, astatine, and iodine. Certain common hand tools were cadmium-plated, and some marking materials (pens and pencils) contained chemicals that caused corrosion. Contaminating elements

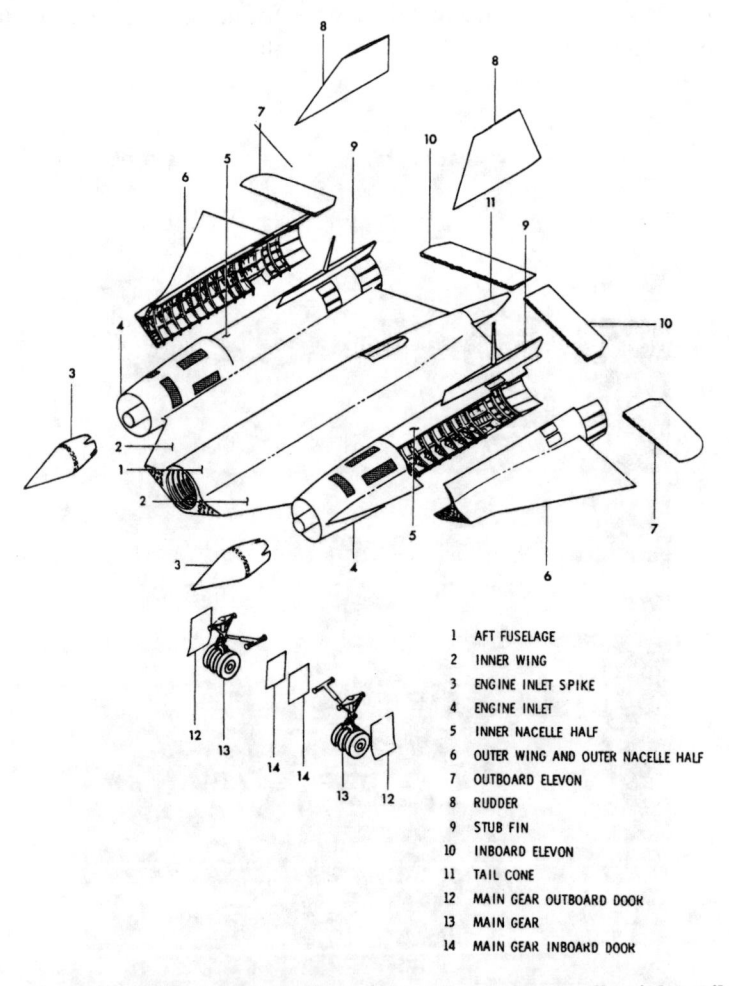

Components of the aft fuselage, including wings, tails, engine nacelles, inlet spikes, and main landing gear. (Courtesy of U.S. Air Force)

also could be found in the composition of solvents, adhesive tapes, paints, packing materials, plastics, fire extinguishing agents, cleaning agents, and other materials commonly used in aircraft maintenance. Technicians and maintenance personnel were required to be vigilant and use only compatible materials from approved lists.[50]

After all of these precautions, Lockheed technicians were baffled when they discovered that wing panels spot welded in the summer failed earlier than expected, but those manufactured in winter lasted indefinitely. The problem was eventually traced to the local city water system; the water supply was heavily chlorinated in the summer to prevent algae growth, but not in winter. Skunk Works personnel solved the problem by switching to distilled water for washing titanium parts.

Technicians discovered a problem with cadmium-plated tools while working on the airplane's engines. Cadmium residue left behind as fasteners were tightened caused microscopic stress-corrosion cracks. When the bolt heads were heated to more than 600°F, well within normal operating temperatures for the engine, they simply fell off.

Prior to drilling and machining titanium alloys, the Skunk Works team conducted a lengthy research program to determine the best tool cutter designs, cutting fluids, and speeds and feeds for the best metal removal rates. Wing extrusions, which were used by the thousands of feet, were particularly troublesome. Initially the cost of machining a rolled mill part was $19 per foot, eventually reduced to $11 per foot as the technique was refined. Engine

The interior of the wing showing corrugations for heat expansion. (Courtesy of NASA)

Lockheed's Blackbird assembly line in Burbank. (Courtesy of Lockheed Martin)

nacelle rings originally consisted of 30 machined parts made on outdated profiling machines and requiring 487 hours to build. A research program costing $1 million resulted in an advanced profiling (shaping) system that produced a single-piece nacelle ring in just 150 hours. Advanced machining techniques combined with Skunk Works research on numerical controls of machining and special tools and fluids ultimately saved the company $19 million during the course of the production program.

Improved manufacturing techniques also were developed at the Skunk Works. To prevent parts from going undergauge while in the acid bath, technicians devised a new series of metal gauges two-thousandths of an inch thicker than standard gauges. Technicians also developed improved drill bits. When the first A-12 was built, a high-speed drill could make only 17 holes before having to be discarded. By the end of the SR-71 program, bits developed at the Skunk Works could drill 100 holes and then be resharpened for further use.

Over the course of 20 years, more than 13 million titanium parts were manufactured at Lockheed. The history of all but the first few can be effectively traced through detailed paperwork back to the original mill pour. For the last 10 million, even the direction of the grain in the sheet from which the part was cut had been recorded.[51]

Lockheed's SR-71 final assembly area in November 1964. These are the first three airframes. (Courtesy of Lockheed Martin)

MATERIALS AND STRUCTURE

Fully 93% of the Blackbird's structural weight consisted of titanium alloys. Because all-titanium construction had not yet become common, Lockheed engineers and technicians pioneered new inspection, test, quality control, and manufacturing techniques. Lockheed technicians found that the difficulty of machining titanium had a great effect on overall construction costs. The initial rate of metal removal from high-strength titanium alloys was only five percent of what the rate would be for aluminum. It was impossible to obtain die forgings to final dimensions or extrusions in finished form. Lockheed technicians had to invent new drills, cutting machinery, powerheads for profilers, and cutting lubricants to increase the rate of metal removal. On some large components, cut on tape-controlled profilers, approximately 90% of the forging weight had to be removed by machining. To save on structural weight, many assemblies contained large numbers of small parts. In similar aluminum assemblies, many of these parts would have been combined to reduce the parts count.

Moreover, large sections of the leading and trailing edges, vertical stabilizers, chines, and inlet spikes were made of "plastic" laminates of phenyl silane, silicone-asbestos, and Fiberglass. These materials—featured primarily on the A-12 and SR-71 families—helped reduce the aircraft's radar signature.

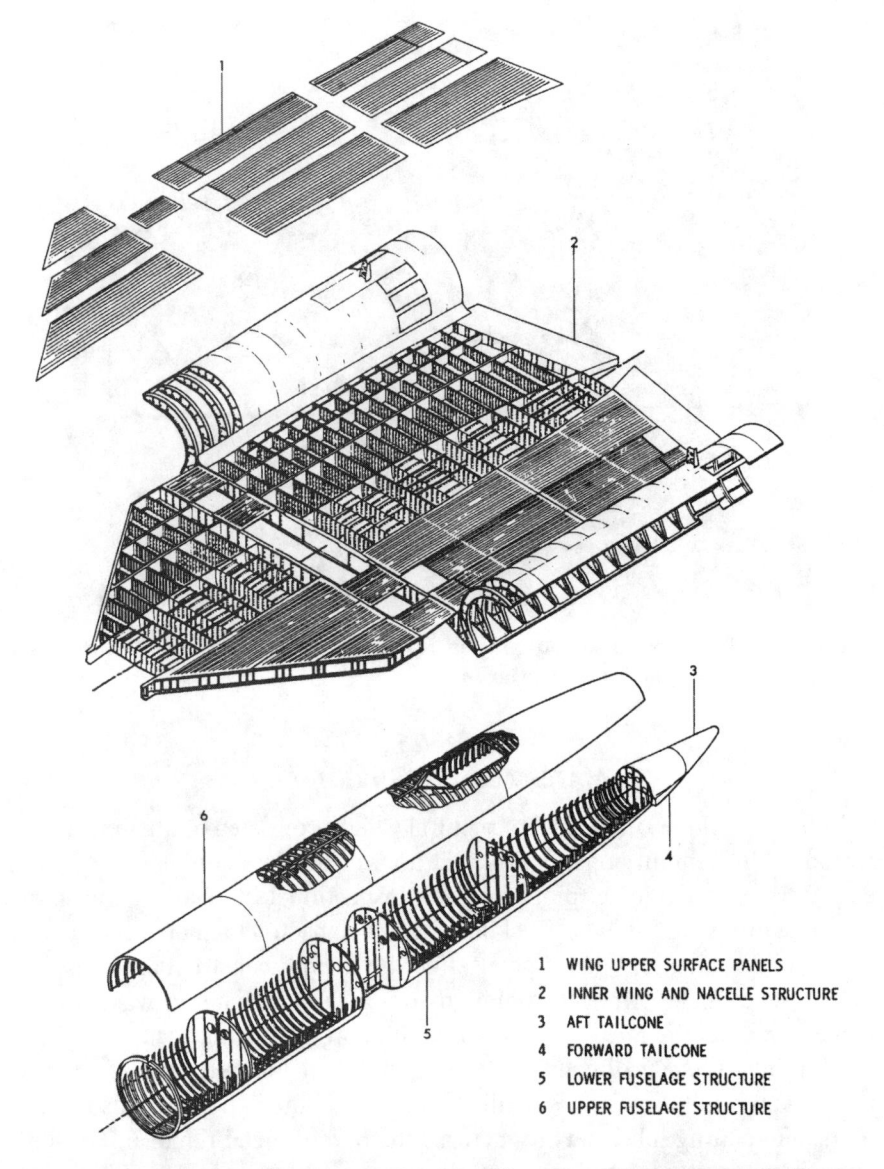

1 WING UPPER SURFACE PANELS
2 INNER WING AND NACELLE STRUCTURE
3 AFT TAILCONE
4 FORWARD TAILCONE
5 LOWER FUSELAGE STRUCTURE
6 UPPER FUSELAGE STRUCTURE

The aft fuselage structure comprised titanium alloy ribs and stringers with titanium sheet skin panels. Upper panels were removable to allow access to the fuel tanks. (Courtesy of U.S. Air Force)

However, the A-12 prototype, A-12T trainer, both M-21 motherships, and the three YF-12A prototypes employed titanium in place of the composite laminates in most areas.

The Blackbirds earned their nickname because they were coated with a high-emissivity black paint to better radiate heat, thus reducing thermal stresses on the airframe. The first A-12 initially flew unpainted. Early models

NECESSITY IS THE MOTHER OF INVENTION 67

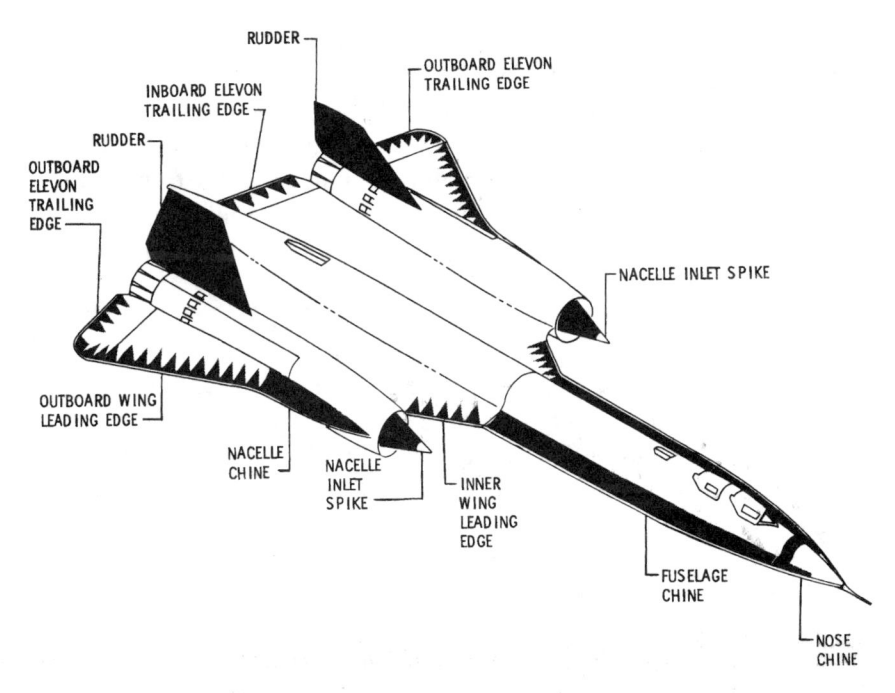

For RCS reduction, most peripheral assemblies of the A-12 and SR-71 were made of composite plastic laminates. (Courtesy of U.S. Air Force)

in the A-12 and YF-12A series were subsequently painted black only on the periphery of the airframe where heating was greatest: on chines, leading and trailing edges, and rudders.

Engineers soon realized it would be advantageous to take advantage of Kirchoff's Law of Radiation that describes how a good heat absorber, such as a blackbody (any extremely dark object), is also an efficient heat emitter. Although convective heating decreases with increasing altitude, heat radiation is independent of altitude. Therefore, in 1964, Skunk Works engineers

This inflight view of the third A-12 clearly shows the composite chine and edge treatments as darker panels contrasted against the bare metal of the airplane's main structure. (Courtesy of Lockheed Martin)

decided to take advantage of the blackbody radiation phenomenon by painting the A-12 fleet completely black.

Eventually, an overall black paint scheme was adopted for all variants. The paint used on the Blackbirds had an emissivity of 0.93 as compared to 0.38 for an unpainted titanium surface. The new paint scheme resulted in a surface-temperature reduction of between 27 and 54°F at cruising altitudes, well worth the additional 60-pound weight penalty imposed by the additional coating.

EXOTIC MATERIALS

Titanium is characteristically light, strong, heat resistant, and nonmagnetic. Its strength compares closely with corrosion-resistant steel, but with just slightly over half the density. Three types of titanium alloys were used in the Blackbirds. The first, designated A-A110AT, contains approximately 5% aluminum and 2.5% tin. The second, B-120VCA, contains approximately 13% vanadium, 11% chromium, and 3% aluminum. Finally, C-120AV contains approximately 6% aluminum and 4% vanadium. Most of the Blackbirds' titanium skin, ranging in thickness from 0.020 to 0.040 inch, consisted of B-120VCA fastened to the frame by riveting or spot-welding.

Riveting is the joining of two metals by means of metal fasteners that mechanically lock into position. This method of is especially applicable to the joining of highly stressed parts, the forming of a discontinuous joint, and in cases where the work is accessible from only one side.

On the Blackbirds, titanium rivets were used in place of more commonly used Monel alloy fasteners that were specified by industry standards at the

The last A-12, Article 133, under construction in March 1964. (Courtesy of Lockheed Martin)

NECESSITY IS THE MOTHER OF INVENTION

time. Rivets are cold driven, and rivet holes require the maintenance of close tolerance to ensure good gripping.

Spot welding, also known as a resistance welding or pressure welding, is a process wherein heat is obtained by the resistance of the metal to the flow of electric current. In spot welding the heat is restricted to a fairly small portion of the lapped area of the parts to be joined.

The Blackbirds also incorporated A-126 corrosion-resistant steel in some parts of the structure and surface panels. This heat-treatable alloy contains approximately 15% chromium, 26% nickel, 2% titanium, and 1% molybdenum. It was capable of withstanding 1200°F, well within the aircraft's performance envelope.

Areas subject to extremely high temperatures, such as the engine nacelle exhaust ejector section, incorporate two types of nickel alloys. Rene' 41 is a

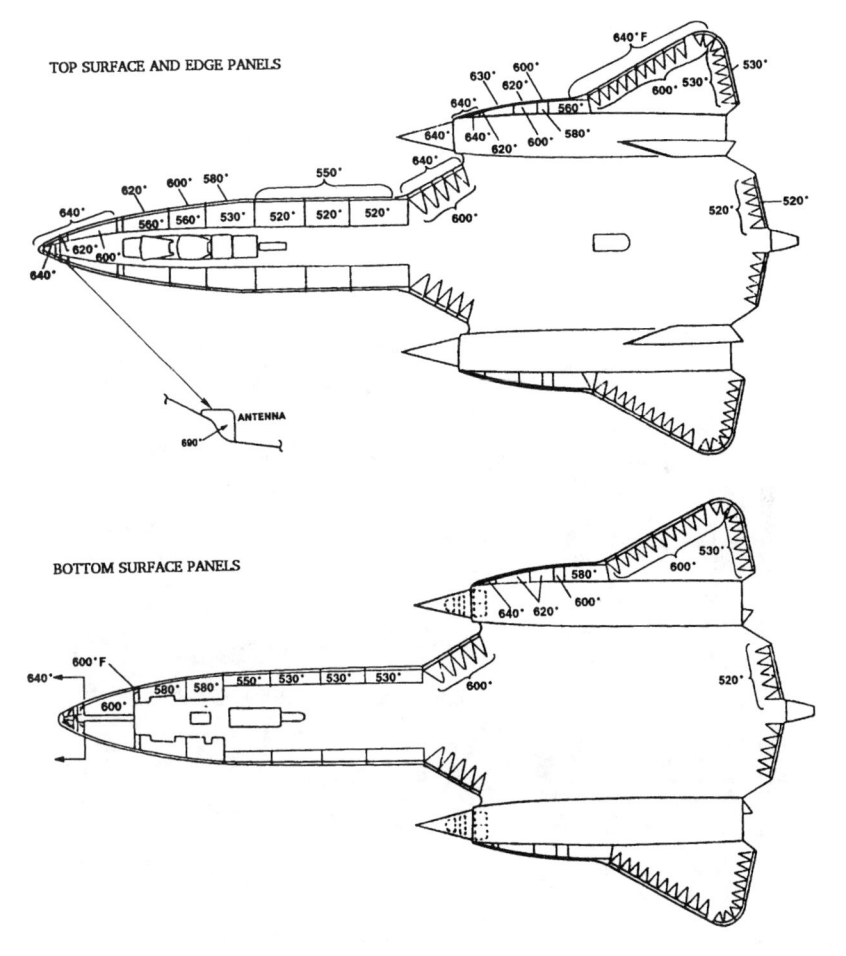

Composite panels on the periphery of the Blackbird's structure had to withstand temperatures in excess of 600°F. (Courtesy of Lockheed Martin)

nickel base metal alloyed with chromium, iron, molybdenum, cobalt, titanium, and aluminum. It can withstand temperatures up to 1600°F. Hastelloy-X—nickel alloyed with chromium, iron, and molybdenum—can withstand approximately 2200°F.

Most of the peripheral assemblies of the A-12 and SR-71 series aircraft were composed of so-called plastic parts, consisting of silicone–asbestos and phenyl silane glass laminates. Designers made extensive use of these materials in the forward fuselage chines, wing edges, inlet spike cone, tail-cone, and vertical stabilizers to reduce the airplane's radar cross section. Composite honeycomb sandwich skin panels, some more than 1 inch thick, were fastened to the underlying titanium framework and were easily removed for maintenance or replacement. These were applied to areas that typically experienced 400 to 750°F during high-speed cruise. Not all Blackbirds featured such extensive amounts of plastic components. On the A-12 prototype, A-12T, M-21, and YF-12A models, silicone–asbestos panels were replaced with A-110AT titanium-alloy skin supported by hat-section stiffeners.

Structural Features

The Blackbirds not only incorporated cutting-edge materials but also some novel design concepts. The Lockheed team developed a monocoque structure for the fuselage and nacelles and a multispar/multirib wing structure

SR-71 forward fuselages were assembled separately from the rest of the airframe. In this photograph, an inlet spike assembly is visible between two fuselage sections. (Courtesy of Lockheed Martin)

with chordwise corrugations for stiffness and to prevent warping at high temperatures. This resulted in a fail-safe redundant structure.

The presence of fuselage side-fairings, or chines, generated nearly 20% of the aircraft's total lift. Acting as fixed canards, they produced a favorable effect on trim drag and minimized the aft shift of the aerodynamic center of pressure as the aircraft's speed increased from subsonic to supersonic. Additionally, vortices from the chines improved directional stability of the aircraft as angle of attack increased. The chines also provided a convenient housing for wires and plumbing on either side of the cylindrical centerbody fuel tanks.

The A-12 series featured a fairly flat, sharply tapered, chined nose. The airplane was a single-sensor platform, capable of carrying a camera or radar in the Q-bay. The YF-12A had a large plastic-laminate radome with a circular cross section to house the fire-control radar for the interceptor's missile launch system. Fuselage chines on the YF-12A ended abruptly at the nose break. The airplane carried no reconnaissance sensors, but was fitted with missile launch bays in the forebody chines. The SR-71 used three interchangeable chined noses for the capability reconnaissance (CAPRE) side-looking radar, optical bar camera (OBC), and advanced synthetic aperture radar system (ASARS). The CAPRE and OBC nose sections had silicone–asbestos chine panels,

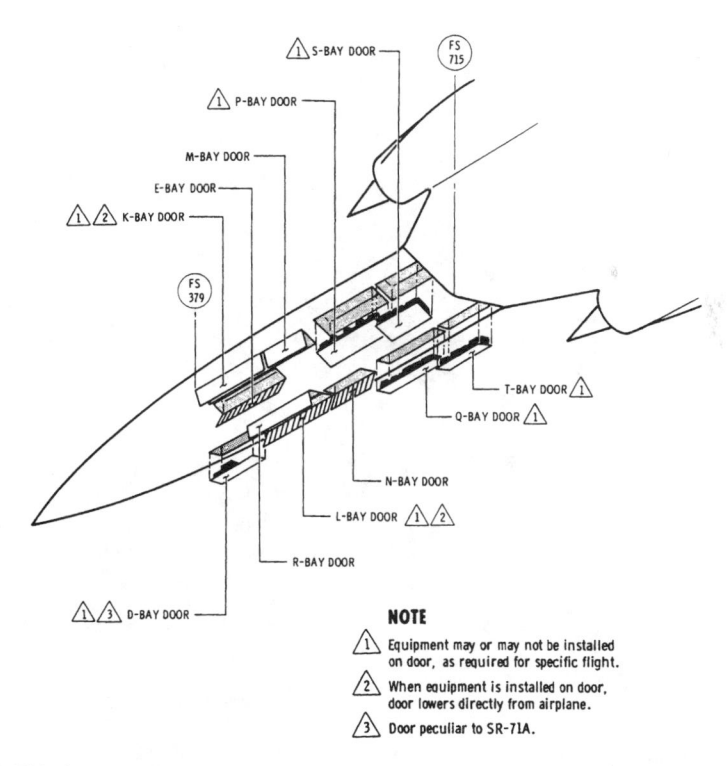

The SR-71's forward fuselage housed numerous equipment bays, mainly in the chine areas. (Courtesy of U.S. Air Force)

whereas the ASARS nose had a one-piece quartz/polymide radome/chine section. The SR-71 was a multisensor platform, capable of carrying a variety of cameras, radar and other mission equipment simultaneously in the nose and fuselage bays.

The forward fuselage primary structure had a circular cross section incorporating rings composed of aged B-120VCA titanium alloy. Tapered chines blended into the sides of the fuselage. The chine structure was not integral with the fuselage structure, but was attached to it as fairings. The chines were partitioned into compartments to house electronics and mission equipment. The fuselage portion was covered with titanium skin, while silicone–asbestos panels covered the chines. The chine support structure was made mostly of annealed B-120VCA. Equipment bay doors were constructed using A-110AT material with some extruded sections as stiffeners. Fuselage longerons, located at the top, bottom, and sides, consisted largely of C-120AV aged-titanium extrusions.

Canopies enclosed each flight station. The A-12 was a single-seat aircraft. All other Blackbirds had two cockpits. The rear-seat position served a different

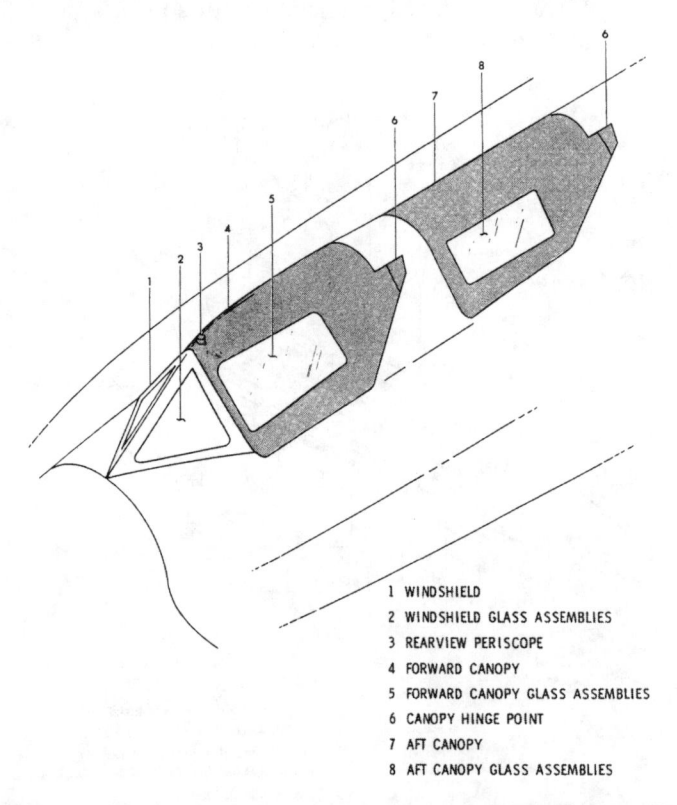

1 WINDSHIELD
2 WINDSHIELD GLASS ASSEMBLIES
3 REARVIEW PERISCOPE
4 FORWARD CANOPY
5 FORWARD CANOPY GLASS ASSEMBLIES
6 CANOPY HINGE POINT
7 AFT CANOPY
8 AFT CANOPY GLASS ASSEMBLIES

Canopies, weighing about 400 pounds apiece, enclosed each flight station. The windows were capable of withstanding high temperatures and high impact forces. (Courtesy of U.S. Air Force)

function, depending on the aircraft model: reconnaissance (A-12, SR-71A), trainer (A-12T, SR-71B/C), interceptor (YF-12A), or mothership (M-21). Each canopy consisted of a titanium frame accommodating two side-glass window assemblies. Each window was composed of two sealed glass panels separated by a 9/32-inch airspace acting as an insulating barrier against aerodynamic heating. Gaskets prevented excessive leakage of cockpit pressurization, but a small amount of air was allowed to bleed through between the panels to prevent fogging. An angular windscreen was provided only for the pilot's position and for the instructor's cockpit on trainer models. It consisted of two glass assemblies, sealed and secured in a V-shaped titanium frame. The windscreen glass assemblies were similar to the side panels, but were coated with magnesium fluoride to reduce glare. All windows were capable of withstanding high temperatures and high impact forces. The laminated inner window assembly consisted of a 1/4-inch-thick outer tempered glass panel, a 1/8-inch-thick silicon plastic layer, and a 3/16-inch-thick inner tempered glass panel. The outer window consisted of a 3/8-inch-thick glass panel. At cruise conditions temperatures reached 420°F on the outer surface of the glass panels and 450°F on the

The pyramidal shape of the airplane's canopies did not allow much headroom for the crew, who wore bulky helmets with their pressure suits. (Courtesy of Lockheed Martin)

adjacent titanium skin. By comparison, boundary-layer air outside the cockpit reached 632°F, and the inner surface of the cockpit was about 80°F. To keep the pilot cool, it was necessary to feed −40°F air into the cockpit to maintain temperatures around 60°F.

The aft fuselage main structure also had a circular cross section and consisted of longerons, bulkheads, rig-frames, and stressed skins. This part of the fuselage was fastened onto the wing structure above and below wing beams extending through the fuselage. Longerons were constructed of titanium alloy and varied in cross section according to load-capacity requirements. Bulkheads, separating fuel tanks and wheel well compartments were of conventional web-and-stiffener design. The bulkheads were made primarily of titanium-alloy sheet and formed sections, with the use of some extruded sections as stiffeners. Titanium ring frames in the aft fuselage consisted of Z-sections, formed into quarter-circle segments attached to the longerons. Titanium skin was spot welded and riveted in place on the fuselage and panel assemblies. Skin panels were attached to the wing spars with fasteners capable of sliding as the panels expanded and contracted with temperature changes, to prevent buckling. Fillet panels were used to blend the fuselage into the wing assemblies. The YF-12A and SR-71B/C trainer models had two small titanium ventral strakes mounted on the underside of each engine nacelle for additional stability, to offset aerodynamic changes caused by their forward fuselage configurations. In addition, the YF-12A also had a titanium ventral fin near the aft end. The ventral fin was so large that it had to remain folded in a stowed position during takeoff and landing. During a NASA research mission in 1975, a YF-12A lost its main ventral fin during a sideslip maneuver. The incident gave researchers an opportunity to flight test a new material.

This view of the fifth SR-71 undergoing maintenance reveals the remarkably sparse chine structure, as well as the leading edge sawtooth panels and wing fuel tanks. (Courtesy of Lockheed Martin)

Bulkheads separating the wing fuel tanks were of conventional web-and-stiffener design. Skin panels were attached to standoff brackets to mitigate thermal effects. (Courtesy of Lockheed Martin)

Technicians fitted a replacement ventral fin, made of Lockalloy, on the damaged YF-12A. Lockalloy, developed by Lockheed California Company, consisted of 62% beryllium and 38% aluminum. Aircraft designers considered it a promising material for use in constructing high-temperature aircraft structures.

The inner wing assemblies consisted of forward and aft wing boxes on either side of the main wheel well. A leading-edge section was attached to the front of the forward wing box and a trailing-edge section to the rear of the aft wing box. The wing box assemblies consisted mainly of titanium alloys, with

On the YF-12A, two fixed ventral strakes on the nacelles and a single large folding ventral fuselage fin offset aerodynamic changes caused by the aircraft's nose configuration. (Courtesy of Lockheed Martin)

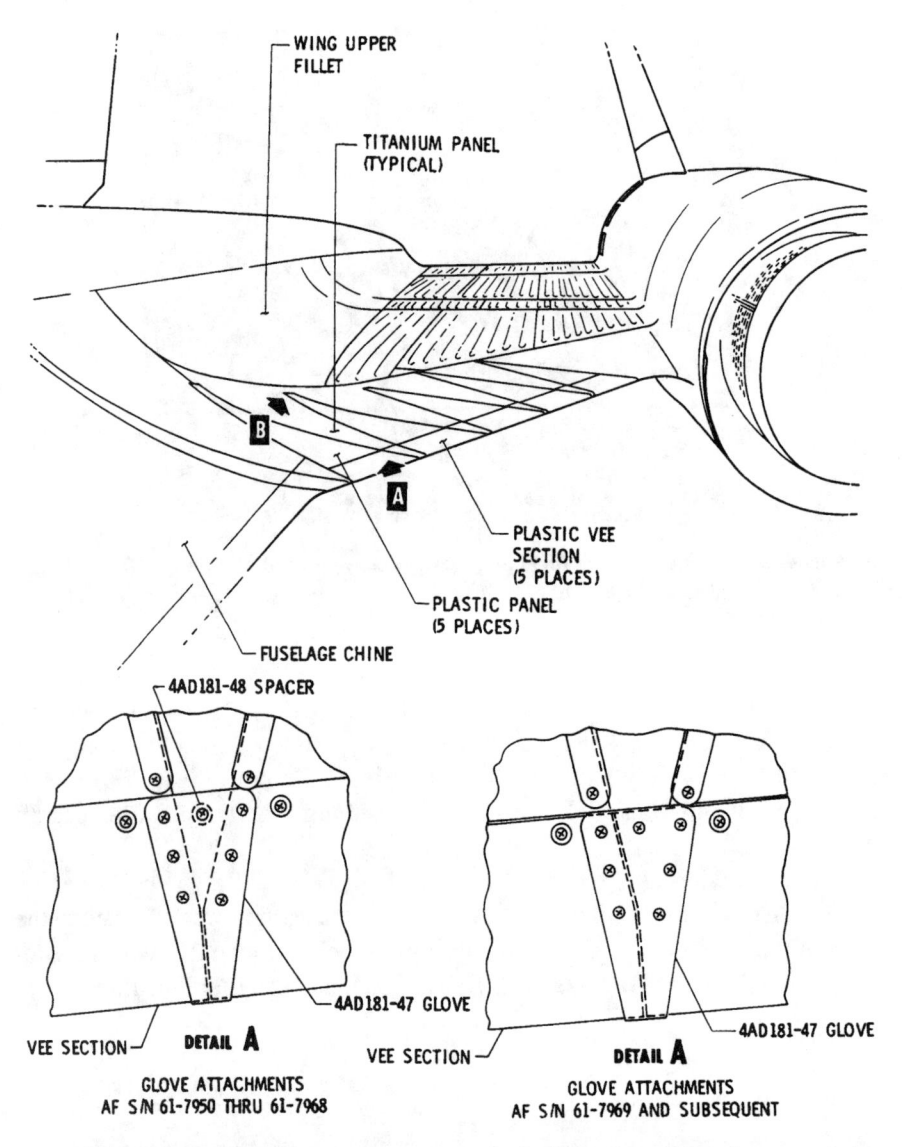

Composite sawtooth (or vee-section) panels on the leading edge of the wings were interlocked with titanium panels of similar shape. This configuration helped reduce radar cross section. (Courtesy of U.S. Air Force)

some corrosion-resistant steel. To reduce the aircraft's radar cross section by reducing radar backscatter, the leading and trailing edges were characterized by triangular plastic panels interlocked with adjacent triangular titanium-alloy panels. The inner wing surface panels consisted of multiple-layer titanium-alloy formed-sheet construction. The lower surface panels were permanently installed and sealed to the wing structure for fuel retention in the integral fuel tanks while the upper panels were removable for maintenance access. The inside surfaces of the panels were formed into corrugations to alleviate effects of aerodynamic heating. Outer surfaces were beaded in the chordwise direction, with beads located between the inner surface corrugations. This type of construction, with the beaded and corrugated portions spot welded together, allowed the panels to expand and contract with changes in temperature.

The outer wing assemblies were built onto the outer half of each engine nacelle. These consisted of titanium-alloy machined forgings and formed parts. Interlocked triangular plastic and titanium panels made up the leading edge.

Inboard and outboard elevons served as the primary control surfaces, hinged at the trailing edge of the wings. The forward section of each elevon was a structural box of titanium-alloy beams, ribs, and skin. The trailing edges were constructed of triangular-shaped titanium-alloy panels, alternated with plastic panels as on the leading edge of the wing.

The two tails were canted inward and mounted atop the aft end of each engine nacelle. Each tail consisted of a stub fin and a rudder. The stub fin was fixed in place, extending approximately 21 inches above the nacelle surface. It was constructed of titanium-alloy machined parts, plate, formed members, and sheet. The stub fin contained rudder servos and housed each rudder pivot post. One full rudder, having no fixed vertical stabilizer, was mounted on each stub fin. Each rudder extended approximately 75 inches above the stub fin. The left and right rudders were identical and interchangeable. The A-12 prototype, A-12T, M-21 motherships, and YF-12A had rudders made from titanium alloy. All others incorporated rudders made largely of plastic materials. The metal rudders were built with a central structural box section and attached leading- and trailing-edge assemblies. Plastic rudders incorporated basic frame members of titanium-alloy. Subordinate members, including some of the ribs, spars, and exterior surface panels, were made of bonded silicone-asbestos reinforced plastic materials. The plastic rudders weighed approximately 500 pounds, and the metal rudders weighed somewhat less.

Each nacelle consisted of an engine inlet, inner and outer nacelle halves, and exhaust ejector. The engine inlet was a barrel-section structure attached to the front beam of the inner wing aft box section. The outboard side of the inlet supported a chine section that faired into the outer wing leading edge. The main body of each nacelle was split, with the inner half built as an integral part of the inner wing. The outer half was hinged for access to the engine. Nacelle ring frames on most of the Blackbirds were spot-welded, built-up, titanium-alloy assemblies. Later SR-71 airframes incorporated machined titanium-alloy forgings in place of most of the built-up ring frames. Longitudinal members of each nacelle half were built up of formed-sheet and extruded titanium-alloy beams. The structure included various bypass doors

Each engine nacelle was split down its centerline, and hinged, for engine access and installation. (Courtesy of Lockheed Martin)

and suck-in doors to control airflow within the inlet and engine compartment. Ejector flaps, operated by differences in air pressure on the outside and inside of each nacelle, were attached to the aft end. Because they were exposed to 1200°F temperatures, the inner surfaces of the ejector flaps consisted of Hastelloy-X. Tracks used to secure the flap sections together were constructed of Rene' 41 alloy with titanium fillers in between each flap.

A movable cone, or spike, controlled the position of the shock wave and the Mach number of airflow within the engine inlet. Bypass doors on the nacelle adjusted airflow depending on pressures within the inlet duct. (Courtesy of NASA)

The engine inlet spike assembly was a conical structure located in the center of each inlet. Moving the spike back and forth controlled the amount of air entering the engine. Spike position was governed by engine air requirements. The spike moved forward during subsonic flight and aft during supersonic cruise. The A-12 prototype, A-12T, M-21, and YF-12A were equipped with titanium-alloy spike assemblies. All other variants employed spike assemblies with a titanium tip and substructure, but external surfaces and some internal components made from silicone-asbestos reinforced plastic materials.[52]

The airplane was assembled in such a way as to alleviate structural loads. The wing bending moment was carried around the engine by frames within the nacelle and through the fuselage by way of continuous spars. Wing surface panels were designed as a beaded (stiffened) structure to carry shear stress but not allow bending. The panels were assembled in such a way as to allow for thermal expansion relative to cooler spar caps. A chordwise corrugation under each skin bead was capable of carrying chordwise loads. Aerodynamic heating affected structural loading as the temperature difference between the top and bottom of the fuselage caused the nose section to droop significantly. Although most of the plastic parts in the chines and wing edges were considered secondary structure, they were required to conform to all local aerodynamic and thermal load limits.

Lockheed technicians attach the forward fuselage assembly to the aft fuselage at the 715 joint. (Courtesy of Lockheed Martin)

The airplane was assembled in two large sections, forward and aft fuselage, and joined together at Fuselage Station 715 (located 715 inches aft of reference Station 0). The aft fuselage consisted of two subassemblies joined at the centerline. In situations involving unusually high loading, as occurred in pitch-up accidents, the "715 joint" was a natural failure point.[53]

ENGINE MATERIALS

Propulsion for the Blackbirds consisted of two Pratt and Whitney JTD-11B-20 (J58) afterburning turbojet engines. Each had nine compressor and two turbine stages. A variable geometry inlet diffuser and a complex bleed bypass system allowed for high engine efficiency in the Mach 2.0 to 3.2 flight regime by controlling the location of the shock wave inside the inlet and allowing air to bypass the turbine section and go directly to the afterburner. The forward compressor stages and inlet case were made of titanium alloys, including Ti-8-1-1 and Ti-5-2.5, because these have good creep (expansion and contraction) properties at temperatures up to 850°F. The first-stage turbine vanes incorporated Mar-M-200DS, a nickel-base alloy that was cast with spanwise columnar crystal grains. Its granular structure reduced the risk of thermal shock cracking. Some first- and second-stage turbine blades, second-stage turbine vanes, and afterburner nozzle flaps were made from another nickel-base alloy, IN-100. The diffuser case was constructed using Inconel-718 nickel alloy, capable of withstanding 1250°F. Most J58 engine components were made of Waspalloy, an oxidation-resistant nickel-base

Technicians work on a J58 engine. The engine casing was constructed from a variety of exotic alloys. (Courtesy of NASA)

With both afterburners at full power during a nighttime ground run, the SR-71 created an impressive display of sound and fury. (Courtesy of NASA)

alloy capable of withstanding 1400°F, but burner components were fashioned from Hastelloy-X. Turbine disks were made of Astralloy, a precipitation-hardened nickel-base alloy suitable for operations up to 1500°F. This extremely expensive material was available as a forging and had creep and tensile strength qualities superior to Waspalloy. Parts with similar applications to Hastelloy-X, but requiring greater resistance to buckling and sliding wear, were made of L-605 (Haynes 25). A cobalt-base alloy L-605 was easy to weld and form. This was later replaced with Haynes 188 and Haynes 230, which had improved oxidation resistance. The exhaust ejector on each engine was supported by streamlined struts and a ring of Rene' 41 on which were hinged free-floating trailing-edge flaps of Hastelloy X. During afterburner operation, these flaps experienced maximum temperatures of 1400°F on their inner surfaces and 1600°F on outer surfaces.

By contrast with some of its other, more advanced concepts, the aircraft operated with fairly conventional flight controls. The inboard and outboard elevons provided pitch and roll, and the two all-moving vertical fins provided lateral control. The vertical control surfaces had to be large in order to counteract the effect of severe yaw during an inlet unstart or engine failure. (An unstart resulted when the shock wave moved outside the engine inlet.)

In addition, the inwardly canted fins minimized roll-yaw coupling with vertical tail deflection and further reduced radar cross section. Internal control linkages included a dual-redundant hydraulic and mechanical system. Because of thermal soak requirements, control cables were made of Elgiloy, a material used in watch springs. Stability augmentation was controlled by a triple-redundant fail-operational electronics system. Although Blackbird designers considered equipping the aircraft with fly-by-wire and adaptive flight controls, they rejected these complex systems because of potential unknown problems that might develop in the extreme operational environment in which the aircraft flew. NASA researchers later adopted an experimental digital flight control system for research purposes. Eventually, in

1981, a digital automatic flight and inlet control system (DAFICS) was incorporated into the operational SR-71 fleet.[54]

Fuels, Fluids, and Sealants

Because it operated in an environment of high aerodynamic heating, the Blackbird required a special low-vapor-pressure, high-flash-point fuel, designated JP-7, one so difficult to ignite that a lit match thrown into a puddle of it is extinguished. Consequently, a pyrophoric igniter called tri-ethylborane (TEB) was injected into the fuel for engine start and afterburner ignition. Fuel for the SR-71—pressurized and rendered inert with nitrogen—was contained in six integral tanks within the fuselage and wing

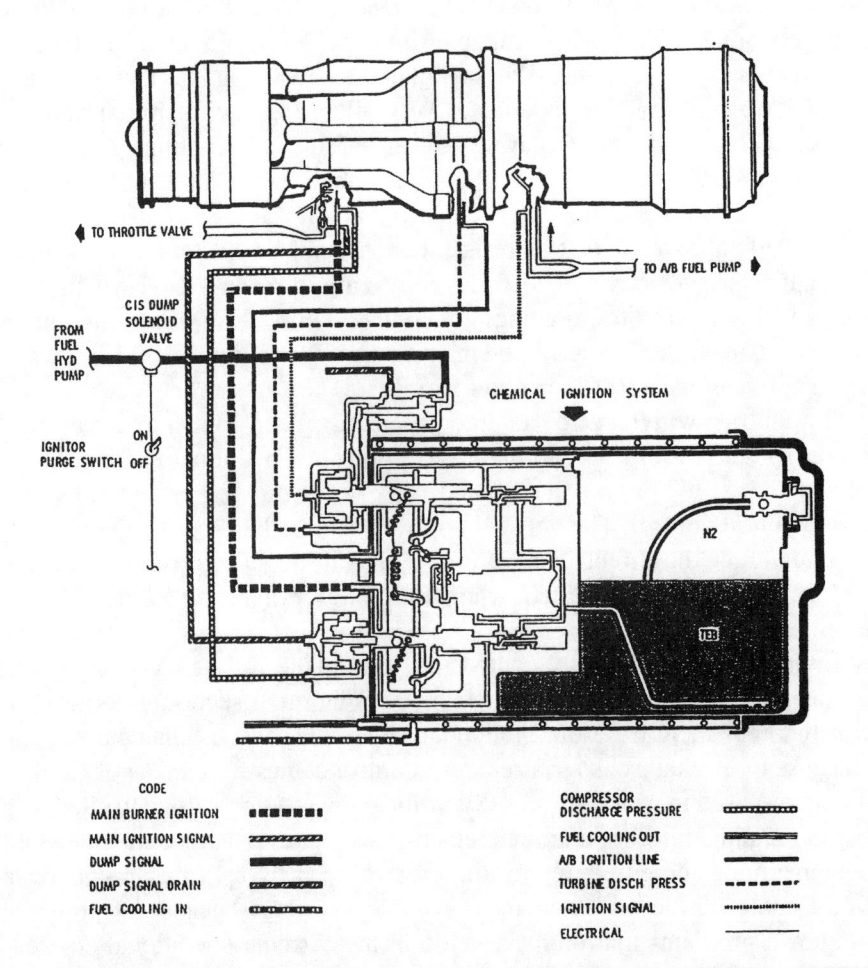

Schematic drawing of the chemical ignition system for the J58 engine. (Courtesy of CIA)

structure. The fuel also served as a heat sink. Before entering the engine, cold fuel was used to precool hot compressor-bleed air for use in the air-conditioning system.

Fuel tank sealing was a major problem for Blackbird designers. Although the tanks were coated with 10,000 linear feet of sealant, they leaked a considerable amount of fuel as a result of the provisions for tank expansion and contraction with changes in temperature ranging from −60°F to more than 600°F. According to NASA YF-12 project engineer Gene Matranga, the sealant material needed to be compatible with titanium, yet "remain elastic enough to move with the expanding and contracting airframe, which grows up to four inches in length when hot, shudders through an unstart, reaches temperatures over 600 degrees and bounces through turbulence and taxi loads." The material chosen was a fluorosilicone that was applied as a liquid using spray guns and paint brushes. It was then cured using special heaters while on the ground. The sealant underwent additional curing by aerodynamic heating of the airframe during flight.[55]

During initial development, a full-scale mock-up provided a realistic simulation of the entire fuel system. It was run for hundreds of hours and simulated various temperature conditions and altitudes up to 100,000 feet. The fuselage angle could be raised to 35 degrees and lowered to dive conditions. Complete fuel gauging, refueling and dump systems were accurately represented and tested. Because the fuel temperature and pressure at the point of final injection into the engine was 600°F and 130 pounds, absolute pressure, the pumping and purging systems required a great deal of development.

Most types of grease with good high-temperature characteristics were essentially solid at normal temperatures. Skunk Works technicians evaluated many types of grease before finding one with satisfactory qualities for lubrication of high-temperature bearings.

Hydraulic oil presented another problem. Johnson's initial queries resulted in one vendor sending him a canvas sack containing a material that would work well at 600°F. Unfortunately, at temperatures below 200°F, it consisted of a white powder and was therefore unsuited to the purpose at hand. Eventually a petroleum-based product developed at Pennsylvania State College was used, but only after Lockheed incorporated a considerable number of additives to improve its lubricating qualities throughout the required temperature range. A full-scale hydraulic system mock-up was also built and tested for hundreds of hours until Johnson was satisfied with its performance.[56]

Seals and pistons presented additional challenges. Because rubber O-ring seals could not be used at high temperatures, stainless-steel rings were used instead. Titanium pistons working inside titanium cylinders tended to gall and seize. The Skunk Works team had to invent chemical coatings to solve the problem.[57]

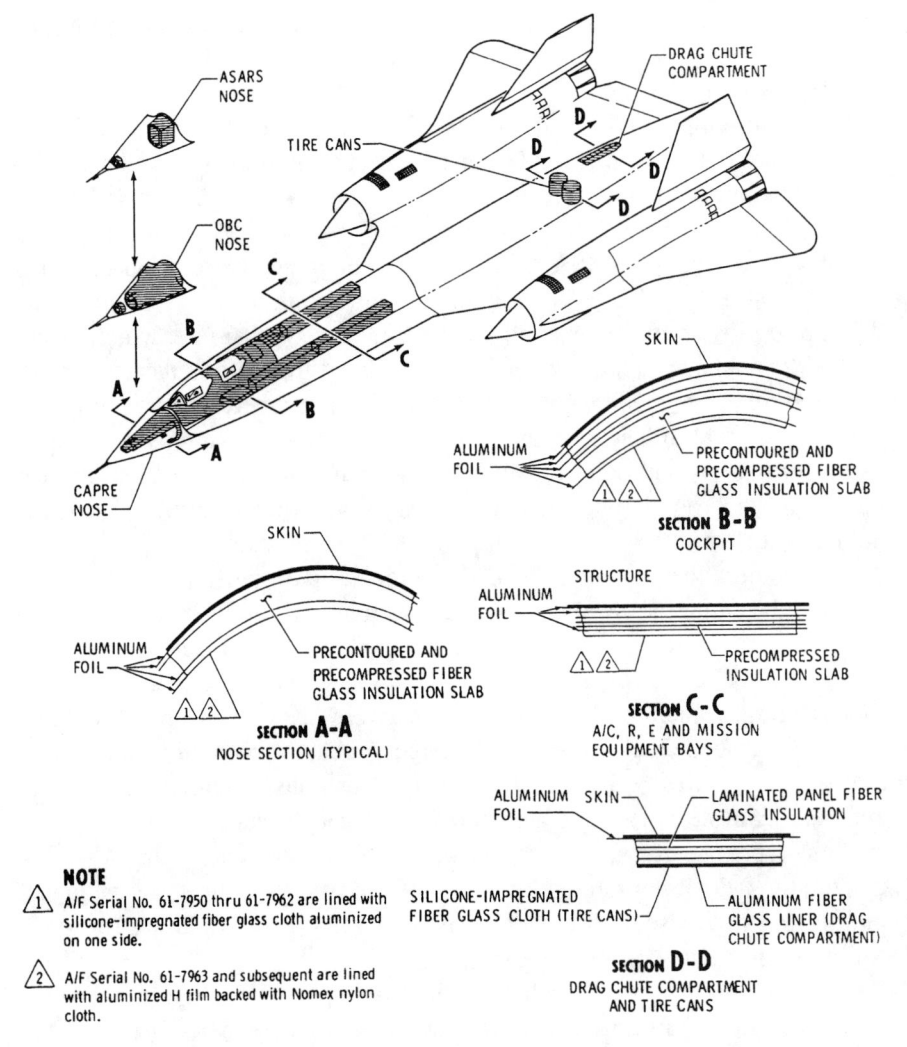

Many areas of the SR-71 airframe had to be insulated against the extreme heat generated during high-speed flight. (Courtesy of U.S. Air Force)

Landing Gear

The landing gear consisted of a standard tricycle configuration with two main struts and a nose strut. These had to support the aircraft during taxi, takeoff and landing, and maintain their integrity while being subject to extreme temperatures.

Each main gear truck included three 22-ply BF Goodrich Silvertown tires, aluminized to reflect heat generated at cruise conditions. These were filled with gaseous nitrogen at 415 pounds per square inch, cost approximately $2300 each, and were good for approximately 15 full-stop landings. A hollow axle allowed the center wheel to be replaced without removing the other two.

NECESSITY IS THE MOTHER OF INVENTION 85

The Blackbird's landing gear had to withstand a wide range of extreme temperature conditions as well as support the airplane during normal taxi, takeoff, and landing operations. (Courtesy of Lockheed Martin)

Each main landing gear included three nitrogen-filled 32-ply tires. Their surfaces were aluminized to reflect heat while retracted into the gear wells. (Courtesy of NASA)

The wheel assemblies were designed so that two tires could carry the limit load in the event of a single-tire blowout. The main gear was mounted between the wing spars such that the wheel and tire assemblies retracted into insulated tire cans. The outer surface of each can was covered with an insulating blanket to retard heat transfer to the tires during cruise conditions. The tire can also minimize damage to wheel-well mounted components, fluid lines, cables, and structure in the event a tire blew out while the gear was retracted.

The nose gear was considerably less exotic. It consisted of two smaller tires without the silver coating. Each nose tire was pressurized to 365 pounds per square inch with nitrogen. The air-conditioning system provided adequate cooling to the nose gear bay.[58]

The main gear had a footprint of 50 square inches per tire. Each main landing-gear tire sustained a load of approximately 17,000 pounds. Each nose-gear tire was subjected to about 9800 pounds. These factors had to be taken into consideration when determining whether a taxiway or runway could sustain the aircraft's weight.[59]

Chapter 4

ABOVE AND BEYOND: BLACKBIRD PERFORMANCE CHARACTERISTICS

The defining performance characteristics of the Blackbirds included flight at high speeds and high altitudes as a result of a unique aerodynamic design and propulsion system. The aircraft operated within an exceptionally large Mach and altitude envelope, but the equivalent airspeed, angle of attack (AOA), and load-factor envelope were relatively narrow. Takeoff and landing speeds were about 210 and 155 knots, respectively. The airplane climbed at 400 to 450 knots equivalent airspeed (KEAS) and operated at normal supersonic cruise speeds between 310 and 400 KEAS. All Blackbird variants were designed to obtain maximum cruise performance around Mach 3.2 at altitudes from 74,000 to 85,000 feet. The external configuration, engine air inlet system, powerplant, and fuel sequencing were optimized at Mach 3.2, and the airplane attained true airspeeds near 1850 knots.[60]

AERODYNAMICS

The distinctive shape of the Blackbirds contributed to their performance capabilities. External configuration features affecting flight characteristics included delta wings, fuselage chines, and location of the engine nacelles.

The Blackbird was configured with very thin delta wings with rounded tips and a twist to the leading edge. The airplane had normal delta-wing flight characteristics including a large increase in drag as the airplane approaches its AOA limit. This resulted in very high sink rates at slow speeds. The wings had a positive dihedral effect (i.e., a right yaw produces right roll and vice versa) that diminished at higher Mach numbers, and the airplane had low roll-damping qualities over the entire speed range. The Blackbird was equipped with a stability augmentation system (SAS) to compensate for poor lateral-directional stability. The outboard portion of the wing's leading edge had a negative conical camber that moved the center of lift inboard, to relieve loading on the nacelle carry-through structure. This also improved the maximum lift characteristics of the outboard wing at high angle of attack, and enhanced the airplane's performance during crosswind landings.

The nose of the SR-71 had the most pronounced chines of any member of the Blackbird family. At supersonic speeds, this structure provided extra lift and improved directional stability. (Courtesy of Lockheed Martin)

The Blackbirds had a pronounced chine (blended forward wing body) extending from the leading edge of the wing to the nose (except on the YF-12A, which had a different nose configuration). This chined forebody, accounting for approximately 40% of total aircraft length, improved directional stability with increasing angle of attack at all speeds, especially in the subsonic range. At supersonic speeds the chines also provided a substantial portion of the total lift and eliminated the need for canard surfaces or special nose-up trimming devices.[61]

Automatic fuel tank sequencing shifted the center of gravity (c.g.) aft during acceleration, corresponding with the aft shift of center of lift with increasing Mach number. The system then maintained the c.g. at a relatively constant optimum location during cruise. This had the negative impact of reducing the static longitudinal stability margin, but the SAS compensated to provide satisfactory handling qualities. Additionally, the airplane's thermodynamic heating characteristics dictated that fuel in the wing tanks had to be used first because of the high surface-area-to-volume ratio.

Engineers also discovered an interesting effect as fuel depletion caused differential heating between the upper fuselage and the cooler lower surfaces where fuel remained. Differential expansion of upper and lower fuselage panels caused the chines to be deflected downward, marginally changing their aerodynamic characteristics.

The midspan location of the engines minimized fuselage drag and interference effects. The inboard cant and droop of the nacelles allowed maximum pressure recovery at the engine inlets at normal angle of attack for high-altitude supersonic cruise. This configuration, however, rendered the aircraft sensitive to asymmetric thrust conditions.[62]

HANDLING QUALITIES

The Blackbird's handling qualities evolved from wind-tunnel model tests that verified the lift-to-drag ratios necessary to achieve mission performance and established the airplane's stability and control characteristics. To meet performance requirements, Lockheed designers had to accept a compromise affecting the airplane's inherent stability and control. In exchange for low drag in cruise, the engineers accepted low stability margins. If they had designed the airplane for high pitch stability, large control deflections would have been required for trim, and the resulting trim drag would have compromised mission performance.

In early wind-tunnel testing, the Lockheed team found an aircraft c.g. with acceptable stability margins caused excessive trim drag. Tilting the nose section downward six degrees alleviated the problem. Reducing the amount of trim elevon deflection that was required also reduced the attendant trim drag. More adjustments were made after the first flight-test data became available.[63]

During category I testing, Lockheed engineers determined the nose-up trim required during Mach 3 cruise was higher than predicted, resulting in excessive trim drag. The nose angle was raised 2 degrees, increasing nose lift at supersonic speeds. This reduced the pitch-trim requirements and associated trim drag. The modification was subsequently incorporated throughout the SR-71 fleet.[64]

FLIGHT CONTROL SYSTEM

Flying qualities also were dependent on the Blackbird's control systems. During early testing the SR-71 featured a primary (hydromechanical) flight control system (PFCS), automatic flight control system (AFCS), air-data display system (ADDS), and Mach trim system.[65]

The PFCS relied upon established airplane design practice, including a full-power irreversible flight control system with conventional stick and rudder pedals to operate the main aerodynamic control surfaces: four elevons and two all-moving vertical tails. The elevons were hinge mounted to the tailing edge of each wing, one inboard and one outboard of the engine nacelle. Each tetrahedral rudder was mounted to a fixed stub fin on the upper aft portion of each nacelle and canted inward at a 15-degree angle.

For such an exotic aircraft, the SR-71 had an entirely conventional cockpit. (Courtesy of Lockheed Martin)

Stick motions were transmitted to a mechanical mixer (located near the tail-cone) through a system of cables, pulleys, and tension regulators. The mixer provided the mechanical geometry necessary to combine the pilot's pitch and roll inputs with electrical trim actuator signals to provide elevon output. This output was then transmitted to the control surfaces through a series of pushrods and bell cranks attached to the elevon servoactuators. Pitch and roll feel springs incorporated into the mixer assembly provided the pilot with control feel proportional to the degree of control-surface deflection.[66]

The early SR-71 primary flight control system was based on that previously used in the A-12 and YF-12A. Two independent hydraulic systems provided power to the servoactuators operating the control surfaces. The servos were closed-loop irreversible mechanisms, so that there was no feedback to the pilot's controls. Thus the feel springs were added to provide the pilot with conventional control forces. Flight-test data resulted only in minor changes to the production version.[67]

Low stability margins and aerodynamic damping inherent at high mission altitudes adversely affected the airplane's dynamic response and handling qualities. Lift (from the fuselage chines) forward of the c.g. destabilized the airplane in pitch. The chines, acting as fixed canards, also adversely affected

To make up for low stability margins and aerodynamic damping inherent in high mission altitudes, the Blackbird was equipped with a stability augmentation system. (Courtesy of Lockheed Martin)

handling in sideslip maneuvers at cruise angle of attack. To make up for these deficiencies, Lockheed designers incorporated an automatic flight control system (AFCS) consisting of a triple-redundant stability augmentation system, autopilot, and automatic pitch trim control (Mach-trim) system.[68]

The AFCS received Mach number, Mach rate, altitude, and dynamic pressure from the air-data display system. The flight reference system (FRS) and astrointertial navigation sytem (ANS) provided attitude and heading data to the autopilot. The ANS also provided steering commands while the autopilot was set on automatic-navigation mode.

The stability augmentation system was a three-axis, dual-monitored damper system. Eight rate-sensing gyros and three lateral accelerometers provided data to the SAS to compensate for any divergence from stable flight and augment the aircraft's basic static and dynamic stability. The system was considered an essential part of the flight control system during normal operation throughout the performance envelope. It was normally engaged in all flight conditions and was designed to operate with full authority in the event of single failures of the electronics or servos. In the event of a serious aircraft malfunction, such as an engine failure, the SAS was designed to prevent excessive structural loads and limit altitude excursions, while the pilot responded to the emergency.[69]

The SAS was designed to be fail-safe and to allow a single failure in each axis without degrading the airplane's handling qualities. Even with the SAS completely inoperative, the Blackbird was considered safe and controllable in all flight conditions. An SAS failure simply meant the pilot faced a more challenging workload.

Prior to the first SR-71 flight, Lockheed designers conducted simulations using a functional mock-up ("Iron Bird") using actual automatic-flight-control-system components. The results allowed them to develop the basic control systems and tailor cockpit control forces to provide satisfactory flying qualities as evaluated by qualified pilots. Other tests were conducted in a 6-degree-of-freedom, moving-base simulator at NASA's Ames Research Center at Moffett Field, California. Lockheed test pilots verified the data during flights in the YF-12A and NASA's JF-100C variable-stability research aircraft.[70]

The Blackbird was equipped with a two-channel pitch and roll autopilot for controlling the aircraft's flight path. Pitch and roll axes could be selected independently, and autopilot authority was limited so that a failure would not overstress the aircraft before the pilot could disengage the system. The stability augmentation system was designed to operate whenever the autopilot was engaged. The autopilot interfaced with the astroinertial navigation system and flight reference system for attitude and heading data and with the air-data display system for airspeed data. There was no yaw autopilot.[71]

The air-data display system included the pitot static system, air-data computer (ADC), triple display indicator (TDI), pressure altimeter, airspeed/Mach indicator, vertical velocity indicator, and angle-of-attack indicator. A pitot probe on the nose of the aircraft sensed total pressure and provided the pilot with AOA and sideslip data as well. The ADC was a mechanical analog computer for determining altitude, Mach, knots equivalent airspeed, and total airspeed from total and static pressures and ambient temperature. The TDI provided the pilot with a digital readout of altitude, Mach, and knots equivalent airspeed.[72]

A Mach trim system imparted artificial speed stability in the Mach 0.2 to 1.5 range. The air-data computer scheduled Mach trim to restore conventional stick forces and trim requirements during acceleration and deceleration, periods during which the autopilot could not be engaged.[73]

Normal Operating Characteristics

Flying the Blackbird was as much art as science. The airplane was designed to accelerate rapidly to rotation speed once the pilot set maximum thrust during takeoff. With pitch-trim set to 0 degrees, the pilot needed to apply a stick force of approximately 25 pounds to rotate the nosewheel. Too much force would result in excessive pitch rate. During maximum performance takeoffs, the pilot had to take care not to overrotate and drag the tail-cone on the runway.

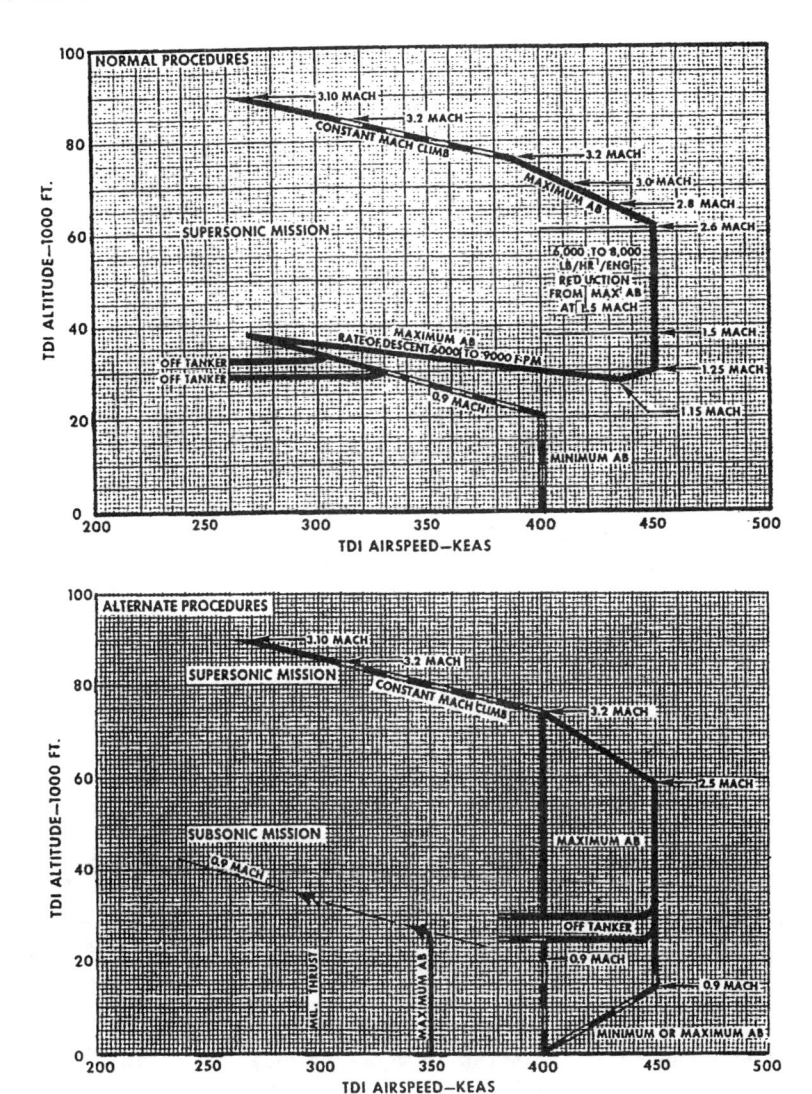

These graphs provide climb speed schedules for the A-12. (Courtesy of CIA)

Climb to cruising altitude involved three phases of operation. These included a subsonic climb, transonic acceleration, and, finally, a supersonic climbing acceleration.

The transonic range was particularly tricky. A light airframe buffet was often felt near Mach 0.90 as airflow conditions changed in the vicinity of the aft bypass doors and exhaust ejector flaps. The pilot observed a Mach jump on his triple display indicator between Mach 0.98 and 1.03 during transition to the supersonic climb speed schedule followed by a decrease in excess thrust from Mach 1.05 to 1.15. He then initiated a shallow dive to improve

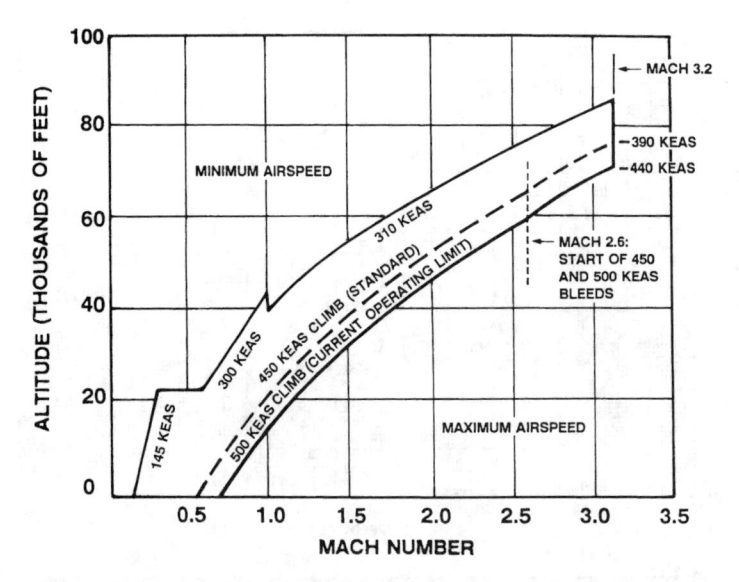

Standard flight envelope for the SR-71. (Courtesy of Lockheed Martin)

acceleration through this speed range. It was best to keep the airplane on a straight course during this maneuver as even gentle turns increased drag sufficiently to decrease acceleration and increase fuel consumption considerably. The pilot had to use great care to avoid overshooting his desired speed during the pull-up.

Once climb airspeed was established, usually at an altitude of about 30,000 feet, the pilot could initiate supersonic climb. It was essential to maintain an accurate inlet schedule to achieve the best possible climb performance. Airspeed control was critical but difficult because the airplane did not respond immediately to small pitch commands. The pilot had to continuously adjust the trim while accelerating to cruise speed.

Occasional episodes of inlet roughness were often encountered between Mach 2.5 and 2.8 and occasionally above Mach 3.0. This phenomenon diminished as the pilot slowly reduced power during transition to cruise altitude and speed.

Depending on mission requirements, planners could choose a flight profile for maximum-range, normal high-altitude cruise, or maximum ceiling. To do this required coordination of three critical interdependent flight parameters: Mach number, equivalent airspeed, and altitude. The selection of values for any two of these variables determined the third. For example, if the mission planner selected a Mach 3.1 cruise schedule with a desired cruise altitude of 73,500 feet, the equivalent airspeed would be 395 knots equivalent airspeed.

Aircraft gross weight, ambient air temperature, and center of gravity also affected performance. Gross weight changed continuously as fuel was consumed (which also affected center of gravity). Because of the airplane's high true airspeed (TAS) at cruise, ambient air temperatures appeared to change abruptly as the Blackbird encountered warmer or cooler air masses. Flight into warmer air caused a decrease in Mach and knots equivalent airspeed while TAS and compressor inlet temperature (CIT) remained constant for a short time. A higher TAS and CIT resulted as the pilot reestablished the desired Mach number. To compensate for effects of variable ambient air temperature, the pilot had to set new cruise altitudes.

Blackbird pilots had to monitor their speed to avoid inadvertently exceeding the desired Mach number. During supersonic cruise at any given gross weight and compressor in let temperature, the altitudes for either maximum range or maximum ceiling flight profiles increased with an increase in Mach number. This increase amounted to approximately 1000 feet per 0.05 Mach number. A small decrease in Mach number (more than 0.05 Mach below desired cruise speed) and knots equivalent airspeed at constant altitude could cause the airplane to reach the ceiling for that speed and become thrust limited. The pilot would then have to descend as much as several thousand feet to reestablish the desired cruise Mach number.[74]

Overall, the Blackbird's handling characteristics were satisfactory. Although stick forces were extremely high at design cruise speed and low lift coefficients, the airplane was usually flown on autopilot under those conditions so that the pilot was unaware of the high forces. The airplane had marginal lateral or directional stability under some conditions, but the stability augmentation system compensated to allow safe maneuvering.[75]

Pilot workload in the A-12 was high because of variations in the turbine inlet temperature that required close monitoring to prevent engine damage. At 7300 rpm unstable temperatures caused thrust variations. The pilot had to trim the fuel flow in small increments with two toggle switches to maintain temperatures within operable limits. Two rotating wafer switches controlled inlet spike position while a third opened and closed the bypass doors. During high-Mach cruise, the pilot had to adjust these switches as needed while determining the optimum inlet schedule of door and spike positions, navigating and maintaining general control of the airplane. Coupled with the autopilot, the stability augmentation system mitigated the problem to some degree.[76]

Propulsion System

Propulsion for the Blackbirds consisted of two Pratt and Whitney JT11D-20 (J58) afterburning turbojet engines with a unique variable-geometry inlet diffuser and a complex air-bleed bypass system that allowed air to bypass the turbine section and go directly to the afterburner, thus acting as a turboramjet. During high-speed flight, the inlet and exhaust ejector generated more than 80% of the total motive force while the engine itself provided

AIRFLOW PATTERNS

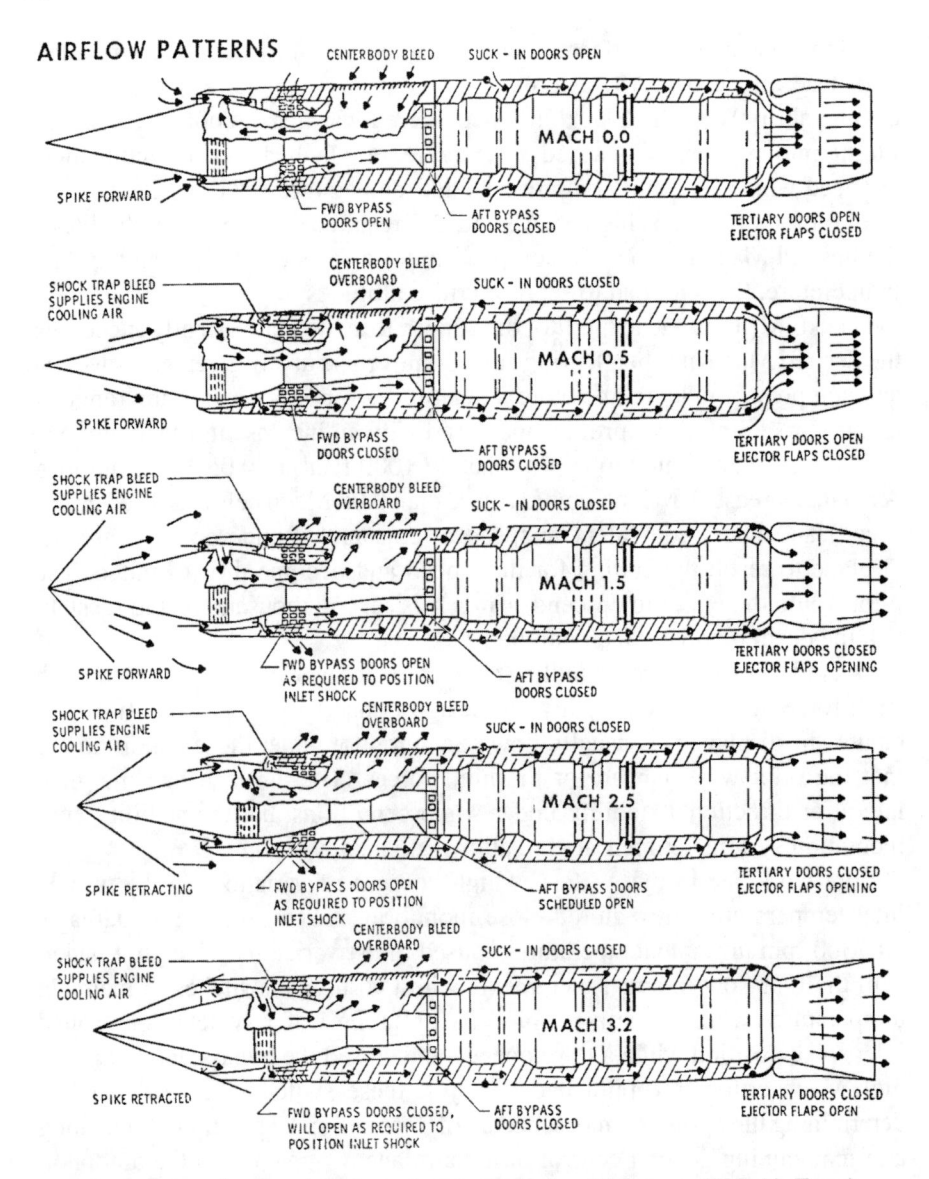

Engine operation from zero to cruise Mach number. (Courtesy of U.S. Air Force)

less than 20%. The Blackbirds had a design cruise speed of Mach 3.2 (about 2100 mph), limited primarily by structural temperature restrictions.

Because there were initially some technical problems with the J58, the first A-12 flights were powered by two Pratt and Whitney J75-P19WSS engines. In July 1962 a single J58 was installed in Article 121, with the proven J75 in the other nacelle, and preparations were made for flight testing. The test crews had difficulty at first because the J58 simply would not start while installed in the airplane. Pratt and Whitney engineers soon discovered a flow reversal causing air that should have flowed out of the compressor's fourth-stage bleed ducts to instead be drawn back into the compressor from the aft end. Lockheed technicians found they could facilitate ground starts by removing an inlet access panel, but this was a stopgap measure at best. Ultimately they solved the problem by cutting holes in the aft section of the nacelle and installing two sets of suck-in doors. Addition of an air-bleed duct also improved airflow through the engine during the startup operation.

Testing of the airplane with the J75/J58 combination continued successfully, and on January 15, 1963, Article 121 made its first flight powered by two J58 engines. The stage was now set for performance envelope expansion flights, but numerous obstacles had yet to be overcome.

The engines' driveshafts began to suffer from wear, twisting and cracking during high-Mach transients. Technicians solved the problem by adding a double universal joint on a new shaft between the gearbox and the engine. Soon the airplane's fuel system began to show signs of fatigue and distortion in the section ahead of the engine. The problem was traced to excessive engine hydraulic pressure feedback caused by large fluid volumes. This led Skunk Works engineers to design a "high-temperature sponge" to reduce pressure to safer levels. Additionally, certain maneuvers resulted in crushed engine plumbing as the outer half of the nacelle pressed against the engine. Pratt and Whitney technicians solved the problem by redesigning the engine mounts and adding a tangential link between the nacelle and the engine to maintain some distance between the two components.

Perfecting airflow in the air induction system necessitated extremely accurate scheduling of the inlet spike's movements and the position of the bypass doors. Fine tuning of these systems gave the pilot greater control over high-speed inlet drag. This reduced the airplane's fuel consumption dramatically. Incremental tests over the course of more than 60 flights gradually expanded the speed envelope from a marginal Mach 2.2 to the design cruise condition. On each flight Lockheed test pilot Lou Schalk gently increased the speed by one-tenth of a Mach number before selecting the next spike progression increment. Afterward engineers analyzed the postflight data and incorporated successful results into the inlet schedule. Trial-and-error testing eventually determined the best combination of spike and bypass door positions. Schalk achieved Mach 3.0 in Article 121 for the first time on July 20, 1963.[77]

The J58 engines had a maximum afterburner thrust rating of 34,000 pounds each at sea-level, standard-day conditions. The Blackbird's propulsion system included a mixed compression inlet in which air entered at supersonic speeds and slowed to subsonic speeds before reaching the engine. The air velocity had to be reduced because no existing engines could run on supersonic flow. Several devices moderated airflow into the engine. A movable cone, or spike, in the inlet translated forward and aft to control the position of the shock wave

POROUS BLEED

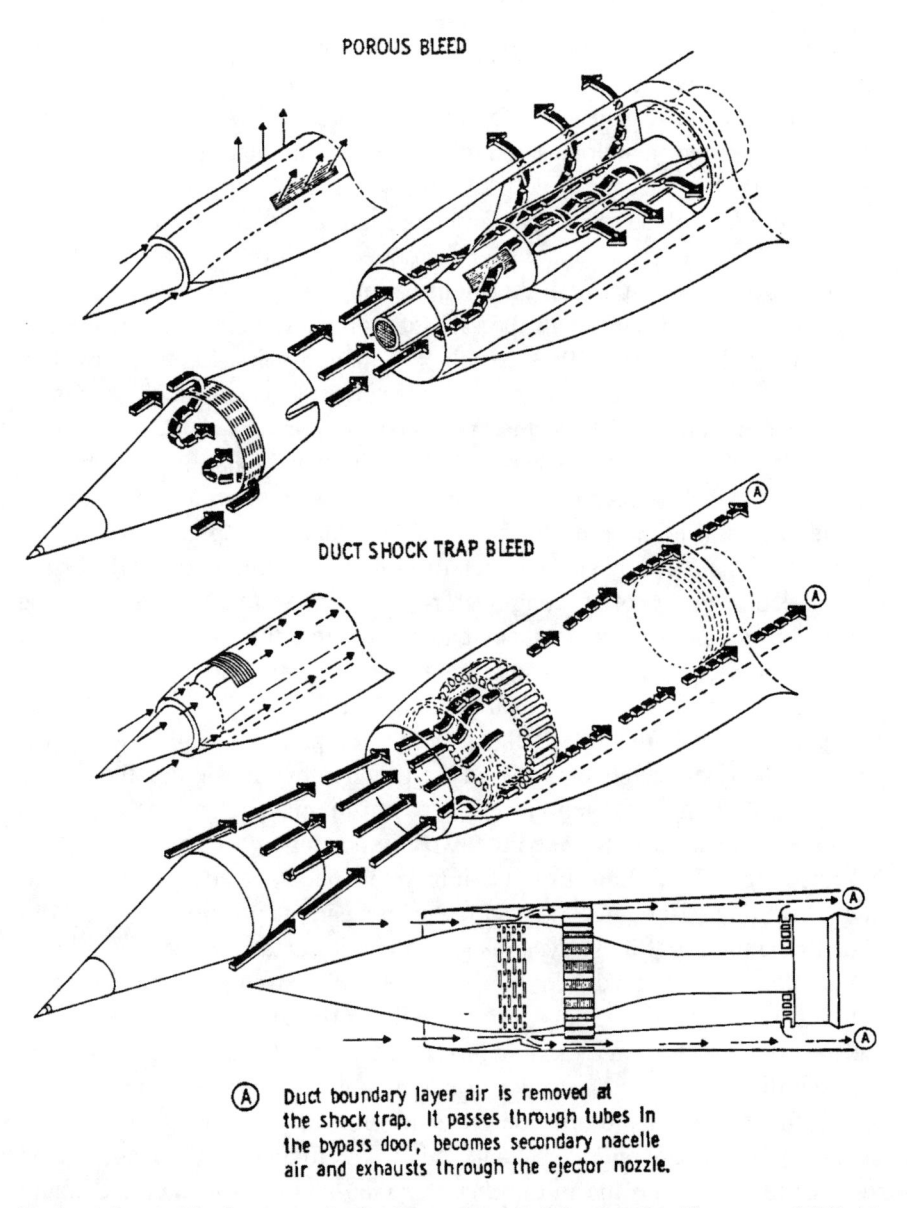

DUCT SHOCK TRAP BLEED

(A) Duct boundary layer air is removed at the shock trap. It passes through tubes in the bypass door, becomes secondary nacelle air and exhausts through the ejector nozzle.

Engine inlet air-bleed patterns in the forward part of the nacelle. (Courtesy of U.S. Air Force)

and inlet Mach number. Forward bypass doors opened and closed to maintain the proper shock-wave position. The doors operated automatically as a function of pressures measured in the ducts. Aft bypass doors, operated by the pilot as a function of Mach number and forward door position, controlled airflow at the engine turbine face. Designers also devised a system to bleed off low-energy boundary-layer air that formed along the surface of the inlet spike.

FORWARD BYPASS

AFT BYPASS

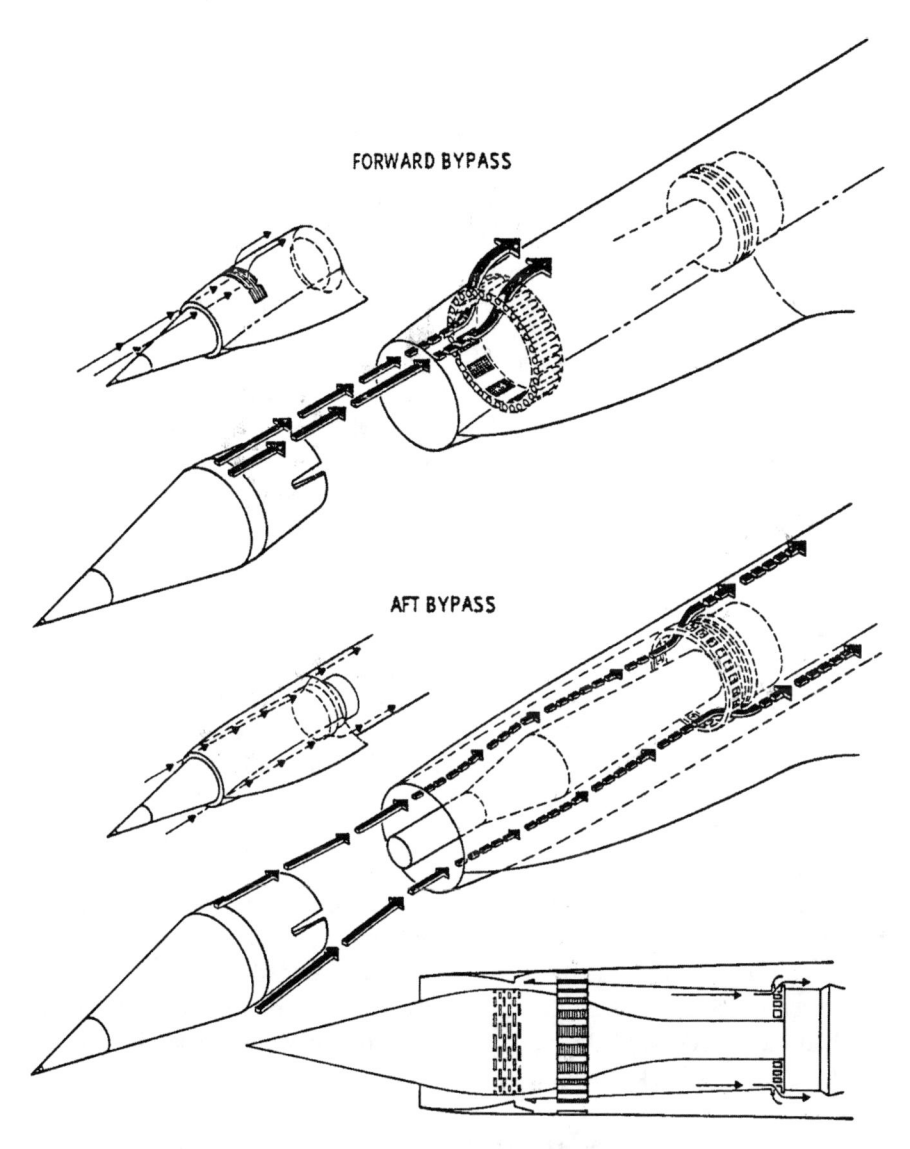

The bypass system directed air from the inlet directly to the afterburner. (Courtesy of U.S. Air Force)

This practice improved inlet efficiency by making the entire main inlet flow passage available to the high-energy, high-velocity airflow.

A CIA report on the A-12 underscored the pivotal function of the Blackbird's inlet:

> A supersonic inlet or air induction system is designed to provide the best possible aerodynamic performance over a range of supersonic Mach

numbers with a stable and steady flow of air to the engine. However, due to constraints imposed by supersonic aerodynamics, truly optimum performance with an ideal shock pattern and an inlet airflow exactly matched to the engine airflow requirement can only be provided at one flight condition. Since the OXCART aircraft must cruise for considerable periods of time at a Mach 3 speed, maximum possible range is realized by providing this optimum inlet performance at the Mach 3 cruise condition. The basic geometry and airflow characteristics of the inlet are then varied to provide a minimum compromise of aerodynamic performance and efficiency at lower flight speeds. Some of this needed flexibility is provided by varying the position of the inlet spike. Since the airflow which can be admitted by the inlet is in excess of that which can be accepted by the engine at other than the design condition, this excess airflow is dumped overboard through a series of forward bypass doors or passed down the nacelle airflow passage around the engine through a series of aft bypass doors.[78]

The inlet system was much more complex than it appeared from the outside. It included the cowl structure, a moving cone-shaped spike to provide optimum internal airflow, a porous center body bleed assembly, and an internal cowl shock trap bleed for internal shock-wave position and boundary-layer flow control. Forward and aft bypass doors regulated airflow inside the inlet and to the engine. Each inlet was canted inboard and downward to align with the local airflow pattern in the vicinity of the engine nacelle. Normally the air inlet control system automatically operated the spike and forward bypass doors, but a manual system was also available.[79]

The spikes regulated the amount of air entering the inlet by moving progressively further aft as Mach number increased. The spike remained fully forward up to Mach 1.4. As the airplane approached Mach 3.0, the spike moved rearward to a maximum of 26 inches. By this point, the inlet capture area had increased by 112% while the throat diameter decreased by 54% in order to hold the shock wave in the correct position.[80]

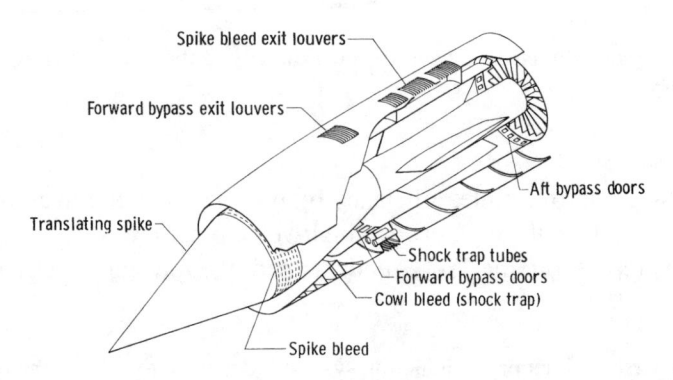

Inlet diagram showing spike assembly and bypass system. (Courtesy of NASA)

For ground operations and flight below 30,000 feet and Mach 1.4, the spikes were locked in the forward position. Above 30,000 feet the spikes were unlocked but remained forward until the airplane reached a speed of Mach 1.6. Spike position was scheduled by the inlet control system as a function of Mach number, angle of attack, and sideslip.[81]

During high-speed flight in the Blackbird, compression of air in the inlets generated most of the vehicle's thrust. At Mach 2.2 the inlet produced 13% of the overall thrust with the engine and exhaust ejector accounting for 73 and 14%, respectively. At Mach 3 cruising speeds the inlet provided 54% of the thrust and the exhaust ejector 29%. At this point the turbojet continued to operate but provided only 17% of the total motive force. The inlet had compression ratio of 40:1 at cruise conditions where each inlet swallowed approximately 100,000 cubic feet of air per second.[82]

A significant percentage of the air entering the inlet bypassed the engine through ducts and traveled directly to the afterburner. At cruise Mach conditions, fuel burned more advantageously in the afterburner than in the main burner section. Hence, engineers described the powerplant as a turbo-ramjet.[83]

Shock waves created by the inlet spikes presented a unique challenge. If designers failed to properly match airflow to the inlet, the shock wave created drag. Normally the shock wave would be expected to occur slightly behind the throat and supersonic diffuser for stability. But in this case, the spike and bypass doors functioned together to retain the shock wave inside the inlet. Sometimes, however, large airflow disturbances or improper inlet control system operation caused the inlet to expel the shock wave. This resulted in an inlet unstart, the byproduct of insufficient pressure and air for normal engine operation. This sudden loss of thrust produced violent yawing,

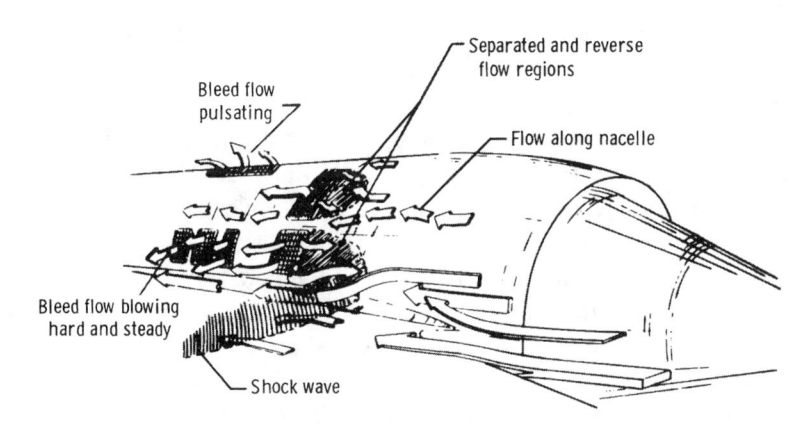

Airflow into the nacelles with forward bypass doors open. Note areas of separated and reverse airflow. (Courtesy of NASA)

pitching, and rolling of the airplane. Pilots likened the phenomenon to "being in a train wreck."[84]

This sudden expulsion of the spike's normal shock wave from the inlet throat caused the airplane to yaw violently in the direction of the unstarted inlet. To recapture the disturbed inlet shock wave, the pilot had to fully open the bypass doors on the unstarted side and move the spike fully forward, reducing the area of the inlet aperture. He then gradually returned the spike and doors to their position prior to unstart.

Lockheed propulsion engineers responded to incessant air induction system problems by changing inlet geometry and speeding up the manual trim schedule to allow quicker acceleration and save fuel during the climb phase. During the first year of testing, the inlet controls were changed frequently, and on many flights the two inlets never seemed to work together properly. Unstarts were a chronic problem until door/spike sequencing was perfected and an inlet control computer was incorporated to operate the system automatically.[85]

The J58 engine was designed to withstand continuous operation at a compressor inlet temperature greater than 752°F. To maximize thrust while avoiding problems associated with high CIT, the single-rotor, nine-stage, 8.8:1 pressure ratio compressor used a bleed bypass system at cruise Mach conditions. Bypass valves opened to bleed air from the fourth stage through six ducts routed around the rear compressor stages, combustion section, and turbine. This air was then reintroduced into the turbine exhaust at the front of the afterburner, where it both augmented thrust and cooled the aft end of the nacelle. The fuel control system scheduled transition from standard turbojet thrust to bypass operation as a function of CIT and engine speed.

Bypass operation usually occurred between Mach 1.8 and 2.0 within a CIT range of 185 to 239°F. In addition to mitigating thermal effects, the bleed bypass process greatly reduced airflow pressure throughout the rear compressor stages. This prevented choking of the compressor by high-velocity airflow. Similarly, by reducing airflow mass, it prevented the front stages from stalling as well.[86]

To further minimize the chances of compressor stall at low airspeeds, Pratt and Whitney incorporated moveable inlet guide vanes (IGV) to control airflow into the compressor section. During takeoff and acceleration to intermediate supersonic speeds, the IGV remained in the axial position to optimize airflow into the compressor. As the compressor-inlet-temperature values rose to the point of bypass activation, the IGV rotated to the cambered position to ensure the CIT remained within operational limits. The compressor had a maximum temperature limit of 800°F. Actuation to the cambered position occurred at around Mach 1.9 and was mandatory above Mach 2.0.[87]

Exhaust-gas-temperature (EGT) sensors monitored discharge temperatures from the two-stage turbine. While operating the engine at cruise conditions,

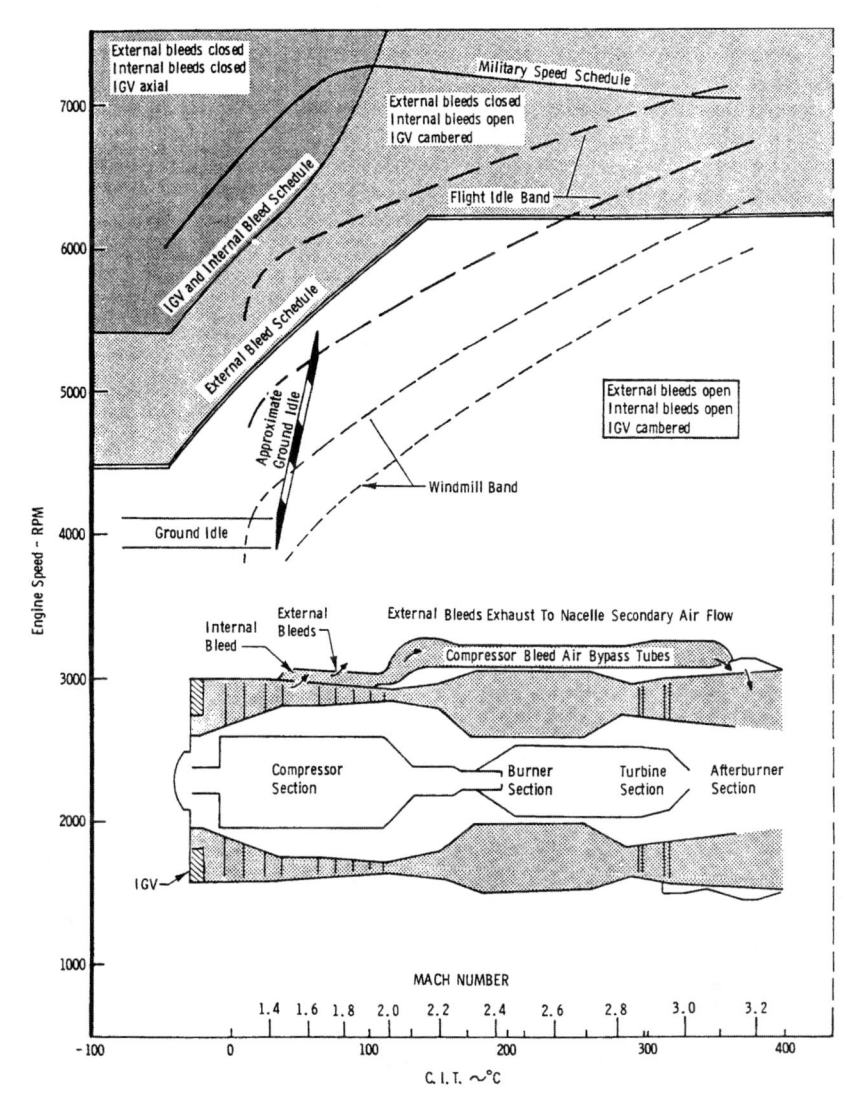

Engine compression bleed and inlet guide vane shift schedule. (Courtesy of U.S. Air Force)

the EGT rose to 2012°F, requiring development of special high-temperature alloys for construction of the turbine section.[88]

The diffuser section diffused air after it passed through the compressor. As the engine's major structural unit, it supported the bearing that accepted all thrust and radial loads from the turbine shaft. The diffuser also supplied high-pressure air for a variety of aircraft functions.

The combustion section consisted of eight cylindrical combustion liners, called cans, arranged in an annular configuration.

A two-stage turbine extracted power from fuel combustion to drive the turbine shaft and compressor blades.

The afterburner was enclosed in a convergent-divergent ejector nozzle. The aft portion had a variable area to regulate backpressure on the turbine and thereby maintain constant engine speed under all conditions scheduled by the main fuel control. Free-floating exhaust-ejector flaps, nicknamed "turkey feathers," opened and closed in response to nozzle pressure changes (a function of Mach number and engine thrust).

Some fuel was diverted for use as hydraulic fluid to power actuators for the inlet guide vane, bleed air system doors, and variable-area exhaust nozzle.

Engine start was usually accomplished using a V-8 engine-driven starter cart connected by a mechanical probe to the engine's main gearbox. The starter turned the compressor while a chemical ignition system injected pyrophoric triethylborane (TEB) into the combustion section and afterburner to ignite the low-vapor-pressure, high-flash-point JP-7 fuel. A TEB reservoir held enough of the chemical for as many as 16 engine or afterburner starts of each powerplant. It could be used during ground operations or in flight to restart a flamed-out engine. During engine start, a green flash signaled ignition of TEB in the afterburner.

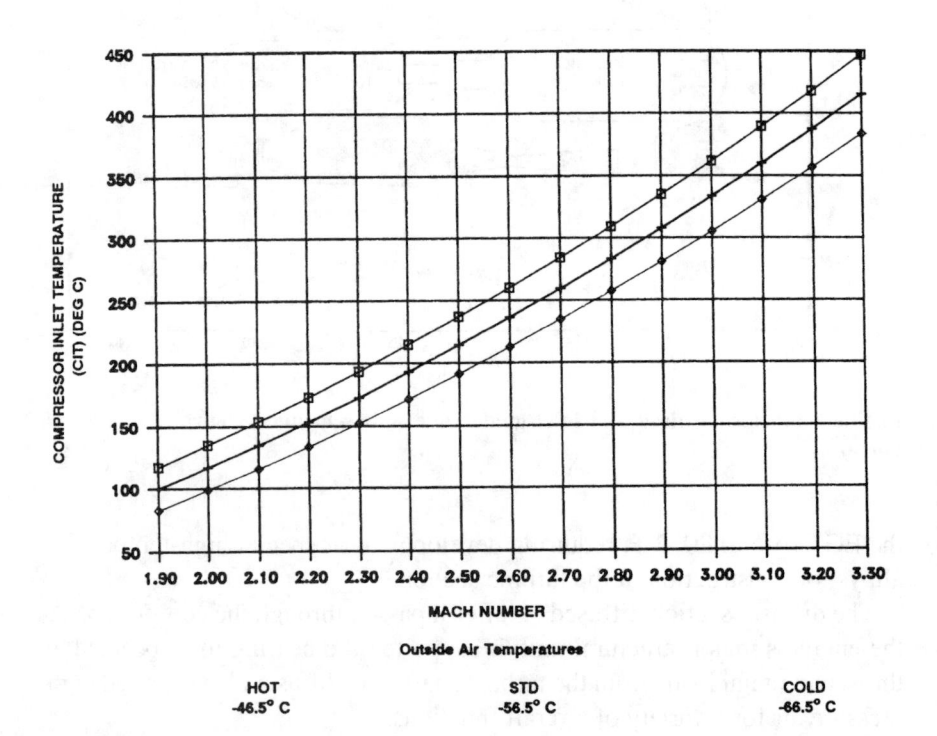

This graph shows the relationship between engine compressor inlet temperature and Mach number. (Courtesy of Lockheed Martin)

During ground operations and low-Mach-number flight, the engine speed varied with throttle position. At higher throttle settings, up to maximum afterburner, the main fuel control scheduled engine speed as a function of compressor inlet temperature and modulated the variable-area exhaust nozzle to maintain constant engine speed. The pilot used a manually armed fuel derich system to protect the engine against dangerous turbine overtemperature by automatically reducing the fuel/air ratio in the burner cans if exhaust gas temperature reached 1580°F.[89]

SPEED AND ALTITUDE

The Blackbirds are best known for their speed and altitude performance. The airplane set numerous records, both official and unofficial. Some of the latter remained unknown to the public for many years.

In 1964, President Lyndon Johnson publicly announced that the Department of Defense had been directed to use the YF-12A to break the official world speed record and inferred the speed would be in the vicinity of 2000 miles per hour. Unfortunately, at the time, the interceptor variant of the Blackbird was the least capable aircraft in the test fleet. It was limited to Mach 2.6 and flying with fixed inlet spikes. The A-12 had been given development priority because it was being readied for operational missions, and, thus far, all flights exceeding Mach 2.95 had been conducted in one aircraft (Article 121).

Although several pilots had accumulated considerable flight time in the aircraft, no realistic operational training missions had been completed. Severe operational limitations were imposed on the aircraft until the test team made further progress on testing the aircraft and mission subsystems. Equipment shortages and deficiencies plagued the program, as well. Hydraulic actuators, afterburner assemblies, and updated inlet controls were in short supply.

Lockheed technicians prepare the third YF-12A (Article 1003) for a test flight. (Courtesy of Lockheed Martin)

Lockheed technicians felt the chances of adequately preparing a YF-12A for a record run before the end of the year were uncertain. Even if the airplane operated perfectly, they felt a 90% probability of success on the first attempt was optimistic. A high degree of precision was required to operate at 80,000 feet and Mach 3 speeds while the piloted attempted to fly the planned corridor for the official measurements.

Air Force, Central Intelligence Agency, and Lockheed officials reviewed their options carefully. The SR-71 was still under construction and would not even make its first flight until the end of the year. The A-12 was still highly classified, and the CIA did not wish to make its existence known. A proposal was made to use one of the two contractor test aircraft (Article 121 or Article 129) under the cover designation "XSR-71" and with a borrowed SR-71 tail number. But whereas this would have increased the chance of success during the record attempt, it would have also imperiled the security of the OXCART program, disrupted testing activities for nearly a month, and delayed the planned A-12 operational capability. Lockheed officials felt they could prepare a YF-12A for speed runs by late October without adversely affecting the other programs. They proposed that practice runs be conducted with the Mach-limited YF-12A while the "XSR-71" option would be held in reserve.[90]

Prior to receiving the speed run orders, preparations were underway to check out and test four A-12 aircraft with an operational capability of Mach 2.8 and 80,000 feet altitude and a range of 2500 nautical miles. Additionally, Article 131 (the mission systems testbed) was to be capable of Mach 3.2, 85,000 feet, and 3000 nautical mile range by January 15, 1965. The Lockheed team continued to struggle with inlet control issues that resulted in high fuel consumption in the transonic speed range, but felt the problems were surmountable. Article 129, with a Lockheed inlet control, demonstrated significantly better performance than those aircraft with an earlier control system, provided by Hamilton Standard.

On August 12, 1964, Lockheed received orders to transfer 37 personnel to Edwards Air Force Base, California, to prepare the YF-12A for the speed run. Engines from Article 121 and several valves from Article 122 were diverted to the YF-12A. According to the orders, National Reconnaissance Office director Brockway McMillan called for making the speed test the highest priority even though it would adversely affect testing and operational capability schedules for the A-12.

Initially, the YF-12A efforts had emphasized armament integration over powerplant integration issues. Now, the most propulsion-deficient airplanes were being pushed ahead on a crash basis in order to achieve the speed record. Kelly Johnson felt this was unwise, but pressed ahead with efforts to modify the aircraft.[91]

As the weeks passed, the situation looked grim, and Director of Central Intelligence John A. McCone asked the agency's Directorate of Science

The YF-12A set several world speed and altitude records on May 1, 1965. (Courtesy of Lockheed Martin)

and Technology to revisit the idea of using Article 121 for the record flight. On 20 August, Col. Jack C. Ledford sent a memo to McCone protesting the idea on the basis of adverse impact to the operational test program schedules and OXCART security. Article 121 was considered a unique testbed because it was equipped with special instrumentation. Ledford felt it should not be diverted from its critical test duties and placed at undue risk for a nonessential project. Additionally, the obviously different configuration would call attention to a member of the Blackbird family previously unknown to the public. By this time, some CIA officials were suggesting the record attempt be scrapped entirely.[92]

Despite these initial difficulties, two of the interceptors at Edwards were eventually ready for the attempt. The president's desire to accomplish this goal before the end of calendar year 1964 was simply unachievable, and the schedule slipped several months.

On May 1, 1965, Air Force crews set several official speed and altitude records in the YF-12A, including a closed-course speed of Mach 3.14 (2070.101 mph) and a sustained altitude of 80,257.65 feet. Although impressive to the public at large, these feats did not represent the Blackbird's maximum capabilities. Just a week later, on May 8, a CIA pilot flew the fifth A-12 (Article 125) to a maximum speed of Mach 3.29 (2171 mph), but this fact remained classified for more than 30 years.[93]

The maximum design cruise speed for all Blackbird variants was Mach 3.2. According to the SR-71 pilot's handbook (flight manual), Mach 3.17 was the maximum recommended cruise speed for normal operations. The pilot, however, could increase speed to Mach 3.3 as long as the engine compressor inlet temperature did not exceed 427°C. Speeds exceeding Mach 3.3 were

This SR-71A set a world absolute speed record of 2193 mph (Mach 3.32) on July 28, 1976. A white cross on the bottom of the aircraft served as a photo reference and aided visual tracking. (Courtesy of Lockheed Martin)

occasionally recorded during test flights, but these operations put excessive thermal stress on the airframe.

Maximum speed was limited by structural temperature restrictions, a part of the flight envelope known as the "heat barrier." Relatively cool outside air temperatures allowed an Air Force crew to set an official speed record in the SR-71A on July 28, 1976, accelerating to Mach 3.32 (2193 mph). This record stood even after the airplane's official retirement flight on March 6, 1990, set a 1998-mile straight-course speed record between Los Angeles and Washington, D.C., in just over 64 minutes at an average speed of 2144 miles per hour. Although it might have been possible to better the speed of the 1976 flight, the crew on the retirement sortie did not want to take that record away from another SR-71 crew.[94]

Designed to fly as high as 90,000 feet, the Blackbirds typically operated between 70,000 and 85,000 feet, altitudes at which they could carry a useful sensor payload and fuel supply. An Air Force crew set an official world altitude record in the SR-71A on July 28, 1976, while cruising at 85,069 feet. Though seemingly impressive, this record already had been broken unofficially during Category II (performance) testing when the fourth SR-71A (Article 2004), with a gross weight of 80,000 pounds, reached a cruising altitude of 86,700 feet. Engineers had predicted the maximum cruising altitude at that weight would be just 85,000 feet.[95]

During one Category II test flight, Lt. Col. Mervin Evenson and Maj. Kenneth Hurley pushed Article 2004 to the upper right-hand (maximum) corner of the performance envelope. After setting up an optimum climb profile, Evenson accelerated the airplane to Mach 3.22 and achieved an altitude of 89,650 feet.[96]

The A-12, a lighter airplane due to its single crew station, was capable of attaining higher operating altitudes than the SR-71. On August 14, 1965, a CIA pilot flew Article 129 to a maximum cruising altitude of 90,000 feet during a test flight.[97]

The SR-71 was approximately 20,000 pounds heavier than the A-12. This meant it would attain altitudes about 3000 feet lower than those attained

This A-12 (Article 129) attained a cruising altitude of 90,000 feet on August 14, 1965. (Courtesy of Lockheed Martin)

by the A-12 at any given point in a flight profile for missions of the same range.[98]

In 1975 Lockheed attempted to determine the feasibility of extending the Blackbird's speed and altitude capabilities. The results of several studies concluded the airplane's maximum speed limit could be extended to Mach 3.5 for

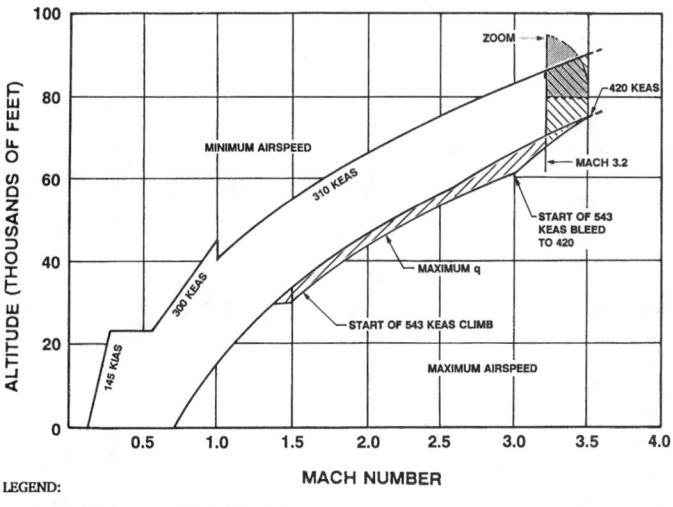

In 1975, Lockheed engineers studied the possibility of expanding the SR-71's performance envelope. (Courtesy of Lockheed Martin)

short periods of time. The only structural limit to speeds above Mach 3.5 was a knots-equivalent-airspeed limit of 420, set by inlet duct pressures and temperatures that exceeded acceptable values. Limited inlet capture area and excessive engine compressor inlet temperatures also limited operations at higher Mach numbers.

Similar studies addressed the possibility of achieving flight in the SR-71 well above 85,000 feet. Results indicated the SR-71 could briefly reach an altitude of about 95,000 feet in a zoom climb profile. The proposed mission could have been accomplished with an airplane at a gross weight of 85,000 pounds. According to the flight profile, the pilot would accelerate from Mach 3.2 to 3.5 at an altitude of 80,000 feet, then zoom to 95,000 feet as speed decreased to normal cruise Mach numbers. The airplane would subsequently settle back down to an altitude of about 84,000 feet. Sustained flight at altitudes above 85,000 feet was limited by wing surface area and engine thrust capabilities.[99]

CRUISE AND CLIMB PERFORMANCE

In conventional jet aircraft, flying faster meant using more fuel. In the Blackbird, however, increased speed resulted in reduced fuel consumption at cruise conditions. For example, the SR-71 flight manual provided specific range charts indicating that an aircraft operating at a gross weight of 100,000 pounds in standard-day temperatures (–69°F) would burn 38,000 pounds of fuel per hour at Mach 3.0. If the pilot accelerated to Mach 3.15, total fuel flow would drop to 36,000 pounds per hour.

During Category II performance testing, the third SR-71A reached an altitude of 89,650 feet while cruising at Mach 3.22. (Courtesy of AFFTC History Office)

Deviations from standard-day temperatures could significantly affect performance, especially during climb and acceleration. As outside air temperature increased above standard-day values, the pilot could feel the Blackbird slow down. The inlets did not function as efficiently because the forward bypass doors opened more, slowing climb rates during acceleration and requiring more thrust during cruise.

Lower-than-standard temperatures improved performance. A fully fueled SR-71 with a 135,000-pound gross weight, accelerating from Mach 1.25 at 30,000 feet to Mach 3.0 at 70,000 feet could burn about 28,000 pounds of fuel if the temperature were 10°C above standard. With a temperature deviation of 10°C (5.6°F) below standard-day conditions, the same airplane would burn only 16,000 pounds of fuel. At cruise conditions a 100,000-pound aircraft could burn 44,000 pounds of fuel per hour if temperatures were 10°C above standard. At 10°C below standard, it would only consume 35,000 pounds per hour.

Because outside air temperature changed along the flight path, the Blackbird pilot had to constantly monitor his compressor-inlet-temperature indicator and fuel consumption. A chart on his checklist compared CIT with Mach number and ambient air temperature, allowing the pilot to recognize temperature deviations as soon as they were encountered. Although the maximum design speed of the airplane was Mach 3.2, the pilot was authorized to accelerate to Mach 3.3 as long as CIT remained at or below 427°C.[100]

RANGE AND ENDURANCE

According to a Lockheed design study, the A-12 had a planned maximum range of 4351 nautical miles, using 60% afterburner, at altitudes between 77,500 and 89,500 feet. The range for a maximum altitude profile (85,500 to

The Blackbirds were equipped to take on fuel from aerial tankers to extend their range. (Courtesy of Lockheed Martin)

97,000 feet) at 100% afterburner was estimated to be 3706. During flight testing, the A-12 demonstrated a maximum unrefueled range of 2800 nautical miles at altitudes between 75,400 and 81,300 feet. A model specification for the SR-71 described the aircraft's maximum range as 3800 nautical miles, using 60% afterburner and an altitude profile from 74,000 to 85,000 feet. Range for a maximum altitude profile of 80,000 to 91,000 feet was 3048 nautical miles.[101]

Operationally, the Blackbirds demonstrated an average range of 2800 nautical miles and could fly extended missions with aerial refueling, limited

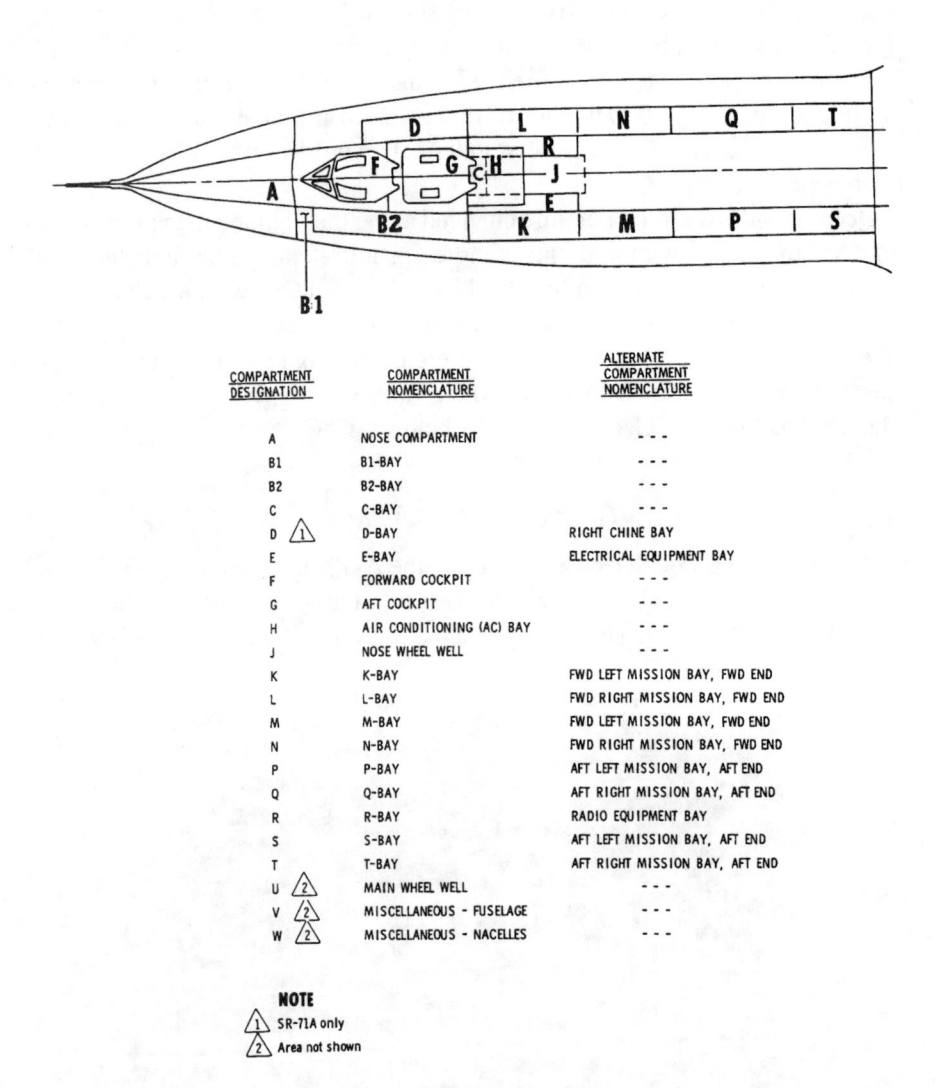

COMPARTMENT DESIGNATION	COMPARTMENT NOMENCLATURE	ALTERNATE COMPARTMENT NOMENCLATURE
A	NOSE COMPARTMENT	- - -
B1	B1-BAY	- - -
B2	B2-BAY	- - -
C	C-BAY	- - -
D [1]	D-BAY	RIGHT CHINE BAY
E	E-BAY	ELECTRICAL EQUIPMENT BAY
F	FORWARD COCKPIT	- - -
G	AFT COCKPIT	- - -
H	AIR CONDITIONING (AC) BAY	- - -
J	NOSE WHEEL WELL	- - -
K	K-BAY	FWD LEFT MISSION BAY, FWD END
L	L-BAY	FWD RIGHT MISSION BAY, FWD END
M	M-BAY	FWD LEFT MISSION BAY, FWD END
N	N-BAY	FWD RIGHT MISSION BAY, FWD END
P	P-BAY	AFT LEFT MISSION BAY, AFT END
Q	Q-BAY	AFT RIGHT MISSION BAY, AFT END
R	R-BAY	RADIO EQUIPMENT BAY
S	S-BAY	AFT LEFT MISSION BAY, AFT END
T	T-BAY	AFT RIGHT MISSION BAY, AFT END
U [2]	MAIN WHEEL WELL	- - -
V [2]	MISCELLANEOUS - FUSELAGE	- - -
W [2]	MISCELLANEOUS - NACELLES	- - -

NOTE
[1] SR-71A only
[2] Area not shown

The SR-71 was capable of carrying 4,000 pounds of equipment in nine separate bays located in the nose, chines, and forward fuselage. (Courtesy of U.S. Air Force)

only by usage of life-support consumables. At a cruise speed of Mach 3.2 and standard-day temperatures, the airplane had a maximum specific range of 54.1 nautical miles per 1000 pounds of fuel.[102]

The range of the airplane was largely dependent on the pilot's ability to endure the claustrophobic conditions of the Blackbird's cockpit for long periods of time. On October 18, 1966, an A-12 (Article 127) demonstrated a maximum endurance flight of 7.67 hours. The mission included subsonic and supersonic cruise and four aerial refuelings. This paved the way for operational sorties and ferry flights to forward deployment areas.[103]

Two months later, Lockheed test pilot Bill Park performed an A-12 proof-of-range sortie with a nonstop 10,200-mile flight in just over six hours. Most A-12 and SR-71 sorties lasted between two and five hours. On April 26, 1971, an Air Force SR-71 crew earned the Mackay Trophy and Harmon Trophy for flying a record-setting 15,000 miles in 10.5 hours. The longest SR-71 operational missions took place during 1987 for the purpose of monitoring the war between Iran and Iraq. Each 11-hour sortie was flown from Kadena, Japan, to the Persian Gulf and back. The longest, at 11.2 hours, took place on July 22.[104]

PAYLOAD CAPACITY AND OPERATIONAL CONSIDERATIONS

Although the A-12 was lighter, it had a much smaller payload capacity than the SR-71. The SR-71 was configured as a multisensor platform capable of carrying integrated photo, radar, infrared, and electronic collection equipment. The A-12 was designed to carry a single major package per mission.

The A-12 featured a 62-cubic-foot central payload bay (Q-bay) behind the cockpit and two 11-cubic-foot chine bays. In all, the A-12 could carry 2500 pounds of mission equipment. The SR-71, on the other hand, was capable of carrying 4000 pounds of equipment in bays located in the nose, chines, and forward fuselage. The total capacity of all of these bays combined was 147.8 cubic feet.

Payload system stability was affected by airframe stability, shock-mount effectiveness, and distortion of bay window glass resulting from uneven heating. Sensors mounted in the nose and on fuselage centerline were more stable than those located in the chines.[105]

Chapter 5

A UNIQUE RESEARCH TOOL: NASA'S MACH 3 FLYING LABORATORY

NASA was involved with development of the Blackbirds from the very beginning. Wind-tunnel tests at NASA Ames Research Center were critical to development of the Blackbird airframe and its unique engine inlet system. During developmental testing at Edwards Air Force Base, the Air Force allowed NASA technicians to install some instrumentation on the SR-71 to collect data, but the agency was unable to obtain a Blackbird of its own until 1969.

The first major NASA project involving the Blackbirds was conducted with the Air Force as a partner. The joint NASA–Air Force YF-12 research program lasted 10 years and produced a wealth of data on materials, structures, loads, heating, aerodynamics, and performance for high-speed aircraft. A second NASA effort in the 1990s employed several SR-71 aircraft as high-speed, high-altitude laboratories to conduct a variety of scientific experiments.

Article 1003 was the first YF-12A to join the NASA–Air Force joint research program. It served primarily as the Air Force test aircraft. (Courtesy of NASA)

Joint NASA/U.S. Air Force YF-12 Flight Research

Within a year of the public debut of the YF-12A, NASA expressed an interest in the aircraft for use as a research platform. In an overview of active and proposed research programs for 1965, planners at the NASA Flight Research Center (FRC, later renamed Dryden Flight Research Center) noted that the YF-12A had "significant features of the configuration and operation of the aircraft that are of vital research interest," and which would complement research being conducted with the XB-70, F-111, and X-15.

NASA engineers regarded the YF-12A, with its capacity to sustain Mach 3 cruise speeds, as a potential source of data for developing advanced supersonic and hypersonic aircraft. Initially, the FRC program consisted of analyzing results of the Air Force–Lockheed test program in hopes of a better understanding of 1) high-altitude hypersonic handling qualities, 2) techniques to determine the structural integrity of hypersonic aircraft in flight, 3) performance of hypersonic airbreathing propulsion systems, and 4) the interrelationships between the aerodynamics of air propulsion systems and the aerodynamics of hypersonic cruise configurations.[106]

NASA officials failed in early attempts to obtain any YF-12A or SR-71 aircraft for flight research. The first request by NASA to R. L. Miller of Lockheed, who in turn submitted a proposal to the company's Director of Flight Test, Larry M. Bohanan, in June 1968, was rejected. NASA engineers had hoped to obtain SR-71 inlet data by installing instrumentation in the fourth service test SR-71A (Article 2004).

Miller agreed to include the NASA request as a separate part of a larger proposal for instrumentation in the aircraft. But even though he forwarded the proposal, Miller did not support it. In the first paragraph he wrote: "It is probably not advantageous from our standpoint to allow NASA participation since it would require increased maintenance and would interfere with our development tests which are required in support of the fleet. In addition, the measurements would not provide any known benefits to the SR-71 program."[107]

Not surprisingly, Lockheed turned down NASA's request. However, a second opportunity presented itself when two NASA representatives participated in the U.S. Air Force Category II tests of the SR-71A. Engineers Gene Matranga and Bill Schweikhard worked with Air Force officials in analyzing propulsion, stability, and control data from the tests conducted in 1968 and 1969.[108]

Although these contacts did not yield an SR-71, the Air Force finally agreed to make two YF-12A aircraft (Article 1002 and Article 1003) available to NASA researchers. A memorandum of understanding (MOU) for a joint Air Force–NASA research program, signed on June 5, 1969, stipulated that the Air Force would provide the airplanes, personnel, ground support equipment, and facilities. NASA, in turn, agreed to pay the operational expenses for the program, using funding that became available following termination of other research programs involving the XB-70 and X-15.

The MOU outlined the general provisions of a joint NASA/Air Force YF-12 research program consisting of Phases I and II. The Air Force Phase I, conducted to explore the tactical performance and support requirements of an advanced interceptor, included tactical tests of command, control, and communications; test intercepts of flying targets; and tests of the ASG-18 fire control system. The program also involved an examination of postattack escape maneuvers, a demonstration of a semi-autonomous operational concept for a Mach 3 interceptor, and an assessment of the feasibility of a visual identification maneuver against an supersonic-transport-type target. The Air Force's renewed interest in Phase I resulted from the recent introduction of the high-performance MiG-25 into the Soviet Air Force inventory. Although Phase I centered on Air Force needs, the memorandum of understanding also accommodated the objectives of the NASA investigations.[109]

The two partners announced the joint program on July 18, 1969. Gene Matranga headed up the NASA team, which spent the first several months of the project installing instrumentation in the YF-12A. By December, engineers had placed strain gauges and thermocouples in the wing and fuselage to measure dynamic loads and temperatures.

The NASA Phase II program began when Paul F. Bikle, NASA Flight Research Center director, signed an agreement on 31 March 1970, with the Air Force, loaning YF-12A (Article 1002) to NASA. This second round of tests included research into propulsion systems, aerothermoelasticity, and flight dynamics characteristics of supersonic cruise aircraft.

NASA YF-12 research represented a cooperative effort by researchers from every NASA aeronautical center. Engineers from Langley Research Center in Virginia concentrated primarily on aerodynamics and structures. Lewis Research Center (now Glenn Research Center), Ohio, had an interest in propulsion aspects. Engineers from Ames Research Center focused on inlet dynamics and the correlation between wind-tunnel and flight data. Researchers at FRC organized these various interests into a single, unified investigation.[110]

According to a program overview by FRC engineer Berwin Kock, "the program also had unique, strong, and continuing support from NASA Headquarters, and the U.S. Air Force was an active partner in the program, providing logistics support and playing an active role in formulating technology experiments."[111]

NASA technicians glued tufts to the fuselage of the YF-12A for surface flow visualization studies. (Courtesy of NASA)

In spite of its earlier resistance to the project, Lockheed Aircraft Company now provided valuable technical support. Privately, however, YF-12 designer Kelly Johnson made some frank observations about NASA and the YF-12 in his personal log:

> Had a visit from the NASA test organizations who discussed their research to date. They haven't come up with anything that is new to us, but it seems to be a good program for them to keep up their technical organizations. I am attaching a letter from Gene Matranga indicating our current relationship, which is excellent. I have two objections to the NASA program, the main one is that they will probably publish important data, which the Russians will be happy to receive as they always are with NASA reports. Secondly, they are repeating so many things we pioneered in and I gravely doubt our people will be given any, or sufficient, credit for solving the problems first. We have continued to cooperate to the hilt with NASA in spite of the above.[112]

NASA flight crew members for the YF-12 aircraft included two pilots and two flight-test engineers. The first NASA crew to be checked out in the YF-12A consisted of Fitzhugh L. Fulton, Jr. (research pilot) and Victor W. Horton (flight-test engineer). In preparation for the project, they underwent a series of three checkout flights in March 1970. Subsequently, Flight Research Center chief research pilot Donald L. Mallick and aerospace engineer William R. "Ray" Young became the second NASA crew.

HEATING AND LOADS RESEARCH

Initially, NASA researchers planned the YF-12 program to focus on propulsion technology, especially inlet performance. Because the YF-12 featured a mixed-compression inlet, engineers planned to investigate drag, compressor face distortion, unstart margins, control parameters, air-data requirements, and bleed system effects.

But problems associated with high-temperature instrumentation delayed the propulsion investigation. This postponement gave NASA engineers time to develop a second initiative: a structures research program involving thermal stresses and aerodynamic loads. The overall effort relied on wind-tunnel data, analytical prediction, and flight research.

Because supersonic aircraft undergo aerodynamic as well as thermal loads, the NASA team planned a series of experiments to measure the both types of loads, combined and separately. Technicians installed instrumentation in the wing and fuselage of a YF-12A (Article 1002). Strain gauges placed in several locations measured aerodynamic loads. At the same time, instruments on the left side of the aircraft recorded skin temperatures.

The airplane enjoyed ideal qualities for thermal research. Previous research aircraft, such as the X-15, had experienced high temperatures but only for short periods of time. The YF-12, however, could sustain high-speed thermal loads for relatively long periods during cruise, enabling temperatures to stabilize. As Flight Research Center project manager Gene Matranga noted,

> We recognized that it would take a while to develop instrumentation for the aircraft, and we decided to use this time to investigate steady-state heating effects on the aircraft structure. In all the X-15 work, everything had been transient. The vehicle went to high speed in a matter of two to three minutes. It slowed down in a matter of three to five minutes. Everything was always transient because the temperature was always increasing or decreasing. The YF-12, on the other hand, could stay at Mach 3 for 15 minutes. We could get steady-state temperature data that would augment the X-15 data immeasurably.[113]

After collecting flight research data over most of the YF-12 performance envelope, researchers compared it to data collected during ground testing in the High Temperature Loads Laboratory (HTLL) at NASA Flight Research Center during 1972 and 1973. The process of comparison involved several steps. Flight research data provided measurements of the combined effects of temperature and loads. Once this information had been gathered, technicians put the aircraft into the HTLL and heated the entire structure to the same temperatures it had experienced in flight. By measuring the strain outputs from temperature alone, NASA engineers could then separate the thermal effects from the flight data to obtain accurate measurement of aerodynamic loads.

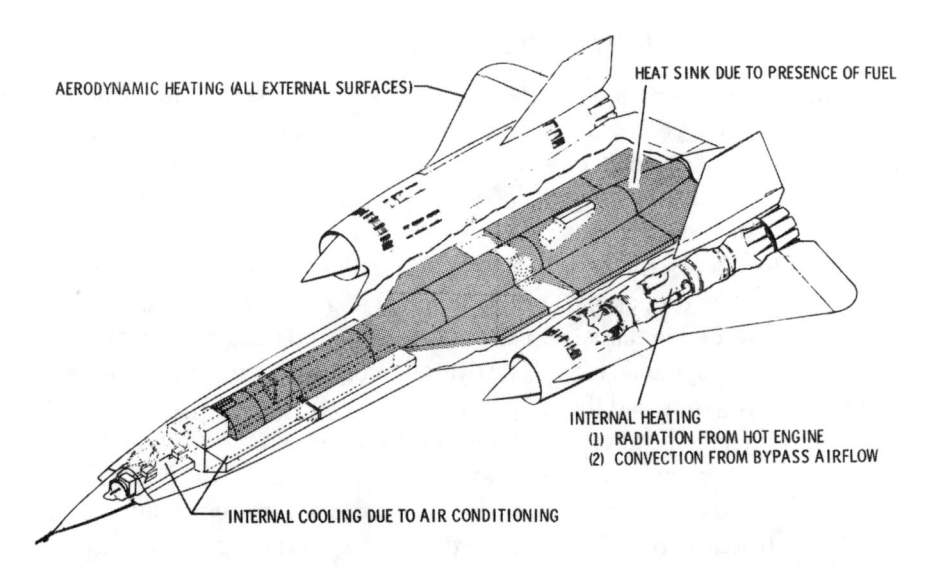

A number of factors contributed to airframe heating and cooling during flight. (Courtesy of Lockheed Martin)

In the High Temperature Loads Laboratory, a radiant heater provided the necessary heating for the ground simulation. The apparatus consisted of 464 stainless-steel reflector panels configured to fit the contours of the aircraft. A total of 16,430 radiant quartz lamps enabled the YF-12 team at the FRC to simulate flight temperatures over a 5000-square-foot area of the aircraft's surface. The heater units covered five areas: aft fuselage, midfuselage and forward nacelles (right and left), and the right and left halves of the forward fuselage. Additionally, another heater fit into the nacelle in place of an engine, allowing researchers to simulate exact three-dimensional engine temperatures inside the nacelle. All of the heaters were subdivided into numerous control zones, each one in turn governed by a surface thermocouple. A data acquisition and control system, a test monitor system, and a test data-processing system fulfilled the remaining simulation requirements. By feeding the temperature profiles recorded in flight into a computer, the quartz lamps could generate the same profiles during the ground tests.[114]

To prepare the aircraft for the heating research, technicians removed the aircraft's vertical tails, nose cone, and inlet spikes, parts not considered relevant to the temperature calibration of the aircraft. They also detached the engines, relying on the nacelle heater to simulate engine temperatures. To avoid accidental explosion, FRC technicians also removed fuel-soaked insulation and replaced it with dry material. They flushed and dried the fuel tanks, as well, and purged them with gaseous nitrogen during the tests.[115]

Results of the heating experiments showed that the predictions largely agreed with the laboratory results. Data obtained during flight, however,

A Unique Research Tool

NASA technicians fitted a heater assembly, equipped with 16,430 radiant quartz lamps, around the YF-12A to simulate flight temperatures. Here, the forebody heater undergoes testing prior to installation. (Courtesy of NASA)

indicated temperatures as much as 20 degrees higher than anticipated because of the differences in the process of heat transfer. The rate of radiant heating is lower than that for aerodynamic heating in areas of higher structural mass. Moreover, the dry fuel tanks used in the ground tests also influenced the results. In flight, the aircraft's fuel acted as a heat sink. Given the absence of fuel in the aircraft during ground-based heating tests, the fuel-tank skin temperatures exceeded those obtained in flight. The simulation and flight measurements converged as the flight-test aircraft depleted its fuel supply. Once these values converged, researchers established a correction for in-flight strain-gauge measurements. With the resultant data about aerodynamic heating at high speeds recorded, the YF-12 team turned to its initial interest, high-speed propulsion research.[116]

Propulsion Research

Using the YF-12, NASA researchers hoped to establish a technology base for the design of an efficient propulsion system for supersonic cruise aircraft, such as a supersonic transport (SST). The main areas under investigation included inlet design analysis, propulsion system steady-state and dynamic performance, inlet engine control systems, and airframe/propulsion interactions. Engineers and scientists from NASA Ames, Langley, Lewis, and Flight Research Center all contributed to the YF-12 propulsion studies.

Because supersonic cruise aircraft require a propulsion system capable of operating efficiently throughout a wide range of flight conditions, designers needed to optimize the inlet system to match engine requirements at varying speeds and altitudes. Dryden engineer James A. Albers described the research process inherent in the YF-12 investigation:

> A first step in optimization of the propulsion system is an analytical study of the various inlet geometries that match the engine requirements. This is followed by wind tunnel testing of scaled models prior to flight-testing. In general, conditions in the wind tunnel do not exactly duplicate flight conditions. With scaled models, the Reynolds numbers and local flow field do not always correspond to those in flight. In addition, the geometry and the instrumentation location and accuracy of wind tunnel models are difficult to match to those of the flight hardware. Since the flight hardware and its expected performance are determined from scaled wind tunnel models, scaling techniques that allow the extrapolation of subscale inlet data to full-scale flight are necessary.[117]

During the YF-12 research program, unscheduled unstarts were common on any given mission. But as a result of the NASA investigation, spike schedule refinements (coordinating spike position to retain the shock wave in the inlet) and hardware improvements rendered unstarts a rare occurrence.[118]

At the same time, NASA did not give up its earlier attempts to acquire an SR-71A for Blackbird flight research. Such efforts did not meet with enthusiasm in Air Force circles. Some Air Force officials felt that sensitive SR-71 technologies might not be protected properly in civilian hands. Indeed, the Blackbirds then in service had improved engines as well as inlet spikes less visible to radar than those on earlier models. For security reasons, the Air Force initially refused to lend NASA an aircraft with either of these advanced features.[119]

Air Force representatives finally relented, agreeing to loan NASA the second SR-71A (Article 2002), but with the earlier model engines, ostensibly to match the inlet behavior of the YF-12A. Additionally, to hide the aircraft's identity, it entered NASA service under the fictitious designation "YF-12C." By May 1971, Lockheed technicians undertook an inspection of the YF-12C in preparation for its addition to the NASA research program. They completed their work on the airplane by the middle of June and prepared it to join Article 1002 and Article 1003.[120]

On June 24, 1971, the program suffered a setback when Article 1003 was lost on its 62nd flight. After performing a handling-qualities evaluation, the crew returned to base. On the way home, disaster stuck. A fuel line in the right engine failed, causing a fire. As the pilot approached Edwards, hoping to make an emergency landing, flames engulfed the right side of the aircraft. The crew ejected safely, but the YF-12A plunged into the desert. The loss of

In 1971, the second SR-71A (Article 2002) joined the program. For reasons of security and politics, it was designated the "YF-12C" and carried a borrowed A-12 tail number. (Courtesy of NASA)

Article 1003 caused delays in the YF-12 propulsion research program. While the remaining YF-12A continued to serve as a loads testbed, the YF-12C arrived at the FRC on 16 July 1971. It did not begin propulsion research flights, however, until June 6, 1972.[121]

Propulsion research using the YF-12C included airspeed calibrations, collection of baseline data, and data collection at numerous flight conditions. To gather data on propulsion system performance, pilots performed such maneuvers as level accelerations and decelerations, constant power turns, and airspeed lag calibration roller-coaster maneuvers. They also gathered data on engine bypass door and inlet spike performance and established speed-power points. Finally, they performed constant-speed climbs and descents at specific knots estimated airspeed (KEAS) or Mach number and constant-power turns. As the crews operated the engine inlet controls in manual and automatic modes, instruments measured oscillations known as phugoids.[122]

For the early Air Force tests, Lockheed had equipped YF-12A with a Honeywell general-purpose computer called the central airborne performance analyzer (CAPA) Phase I to monitor the aircraft's air inlet control system (AICS). The unique inlet system on the YF-12 made it vulnerable to high stresses, severe environmental conditions, and possible malfunctions. Because the AICS only realized its full operational capabilities at high-supersonic speeds, malfunctions—which tended to occur in this regime—proved most difficult to detect. Such malfunctions could not be discovered during static ground tests. But the CAPA provided a central system that would continuously monitor and analyze the performance of the AICS during flight and transmit maintenance messages identifying the faulty components.

Honeywell delivered an improved central airborne performance analyzer Phase II to Lockheed in February 1973. The new CAPA featured a special-purpose computer with an instruction repertoire specifically tailored to the task of in-flight performance monitoring, malfunction detection, and fault isolation. A component called the remote unit monitored additional signals including fuel and hydraulic systems and engine functions for information purposes rather than fault detection. Lockheed technicians installed the CAPA Phase II prototype in a functional mock-up (FMU), a ground-based model containing all of the operational systems of a real YF-12 aircraft. Lockheed delivered the CAPA unit to NASA in March for installation in the YF-12A. By June 28 NASA technicians had completed the integration of the CAPA Phase II system. Between July 12, 1973, and June 6, 1974, the CAPA system operated during 28 flights. Because of the high reliability associated with the air inlet control system, the CAPA made only one valid fault detection/diagnostic in over 71 hours of operation. The remote unit, however, detected failures in other systems, revealed in later ground-based data processing. Had the maintenance software for these systems been installed into the CAPA, technicians on the ground could have detected the failures in real time.[123]

Once this software had been evaluated, the YF-12 crews conducted a series of tests between November 1976 and July 1977 to correlate specific wind-tunnel data points with flight-test data points. To establish the match points precisely, the pilot had to fly the aircraft within 0.05 Mach number of the recorded wind-tunnel speed. The match points provided flight-test inlet data and verified the earlier wind-tunnel results.

Researchers also studied the phenomenon of unstarts by inducing them intentionally and then restarting the engine. NASA technicians installed an inlet override system, called Inlet Recall, to allow manual control of the inlet spike and bypass doors. During a test of the system, the autopilot controlled attitude and altitude. The pilot slowly trimmed the inlets to maximum performance, then went beyond and induced the unstart condition. As the aircraft underwent a sudden, violent loss of altitude and airspeed, the pilot initiated either an automatic or manual restart procedure. These important investigations were complemented by another phase of NASA Blackbird flight research, involving the landing characteristics of these high-speed vehicles.

Landing Studies

In something of a departure for Blackbird researchers, NASA and Lockheed engineers investigated space shuttle landing dynamics using the YF-12C. Several flights, conducted in April and June 1973, demonstrated shuttle-type flight characteristics during low lift-to-drag (L/D) approaches. Specifically, the researchers needed data for L/D ratios of 2 to 3, the range predicted for the space shuttle orbiter. This necessitated operating the YF-12C in a high-drag configuration, achieved by reducing power to idle, moving the inlet spikes forward, and opening the bypass doors to the restart position.

A UNIQUE RESEARCH TOOL

NASA research pilots flew the YF-12A to study landing dynamics of low-aspect-ratio supersonic aircraft. The resulting data validated computer simulation methods developed at NASA Langley Research Center. (Courtesy of NASA)

In addition, the pilots needed to transfer fuel to maintain a forward center of gravity and burn off fuel to allow descent at as light a weight as possible (to avoid flying the aircraft at maximum L/D). The descent profile maximized engine negative thrust-inlet drag and also allowed for the lowest possible lift coefficient.[124]

Three flights, including 26 approaches, resulted in satisfactory pilot ratings for all handling qualities. The flight crews noticed no tendency toward pilot-induced oscillation and suggested that the YF-12C would serve as an acceptable model for space shuttle landing characteristics.

In 1974, NASA engineers decided to use the YF-12 aircraft in a somewhat different role: to study the landing dynamics of a low-aspect-ratio supersonic aircraft. The data would be used to validate the Flexible Aircraft Takeoff and Landing Analysis (FATOLA) computer program developed by NASA Langley, one that offered a six-degree-of-freedom rigid-body simulation. Technicians from McDonnell Douglas Astronautics Corporation programmed FATOLA with the YF-12 structural mode data and computed the airplane response to taxi, landing, and takeoff using a measured runway profile.[125]

Up to this time, aircraft designers lacked a solution to a complex problem affecting transport aircraft. They could not predict adequately the structural and control problems resulting from landing-gear and airframe interactions. Runway irregularities routinely affected tire loads. Surface roughness, ground contour elevations and slopes, and airplane-to-ground axis orientation all contributed inputs through tire deflections and unsprung mass excitations. The increased structural flexibility and higher takeoff and landing speeds of proposed supersonic transport (SST) designs magnified the problem. Because the YF-12 shared many structural characteristics with SST designs, the Blackbirds assumed a leading role in the landing dynamics research program.[126]

Planning for the landing study involved much teamwork inside and outside NASA. Gene Matranga and Jim McKay represented the Flight Research Center. Bob McGehee and Huey Carden were the primary NASA Langley researchers. Bill Fox and Gus Dishman of Lockheed Advanced Development Projects provided additional support, particularly regarding the calibration of instrumentation. In January 1975, after completing these calibrations, Jim McKay submitted a request for project approval to measure YF-12 response to runway roughness. McKay proposed instrumenting the YF-12A to measure ground loads and dynamic response during landing, taxiing, and takeoff on the Edwards Air Force Base main runway.

The program coordinated the research efforts of the NASA Langley Structures and Dynamics Division in developing an active landing-gear control system for proposed supersonic-transport aircraft. Flight Research Center researchers obtained experimental response data from flight tests to correlate with the response calculated using the Flexible Aircraft Takeoff and Landing Analysis program. The validated FATOLA program defined the interactive characteristics of active control landing-gear systems with other aircraft characteristics and systems such as engine thrust, ground effect and crosswind aerodynamics, unsymmetrical touchdown conditions, airframe structural elasticity, and antiskid braking.[127]

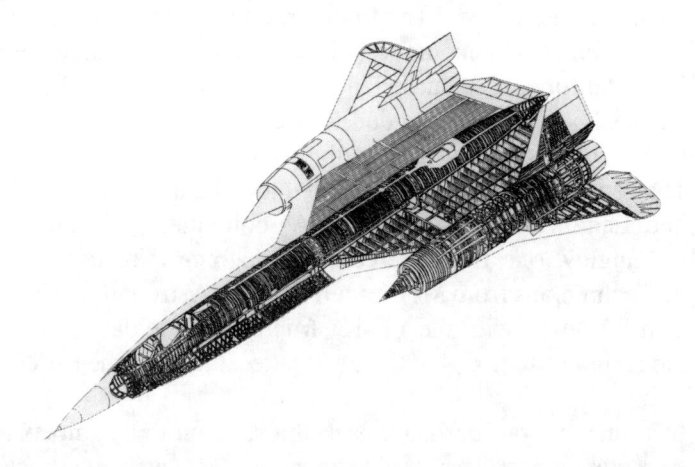

This cutaway view illustrates the complex internal structure of the YF-12A. NASA engineers were interested in the structural dynamics of this highly flexible airframe. (Courtesy of NASA)

During March 1975, Flight Research Center technicians instrumented the YF-12A with strain gauges and accelerometers for the two-phase program. The first phase consisted of consecutive takeoffs and landings in the low-speed flight regime. The second phase included high-speed flight to assess the effect of elevated temperatures on landing-gear performance.

Prior to conducting ground loads and aircraft response experiments, researchers created a runway profile for the primary paved airstrip at Edwards to better define flight-test conditions. A three-track profile, encompassing an area 18 feet on either side of the runway centerline, was then programmed into the FATOLA program.

The first data flight took place on January 27, 1976. Fitz Fulton and Vic Horton conducted a low-speed taxi test and made a number of touch-and-go landings for high-speed taxi data. Don Mallick and Ray Young made a second flight eight days later. Two other flights failed to gather data, once because of foam on the runway and once because of an unsafe landing-gear indication on the YF-12A. Based upon the correlation of the computer simulations and flight data, researchers validated the Flexible Aircraft Takeoff and Landing Analysis program as a versatile analytical tool.

A second landing project took place in 1977. This research demonstrated a dual-mode adaptive landing-gear system to reduce the dynamic response of an airplane during ground taxi. An airplane's landing-gear system absorbs the kinetic energy associated with vertical velocities at touchdown and generally produces maximum efficiency at its maximum sink rate. Designers accomplish this by adjusting the combined shape of the static air load-stroke curve and the hydraulic damping curve to provide a constant load during a maximum-design sink-rate landing. For a given kinetic-energy absorption, this yields the shortest gear stroke. Such a design resulted in the lightest practical gear to absorb the landing energy. An adaptive landing-gear system can also increase the life span of an airframe by reducing vibration stress incurred during taxi, takeoff, and landing.[128]

Such an approach reduced total aircraft weight, but did not necessarily result in a static air load-stroke curve suitable for isolation of runway roughness during taxi. This particular condition had special significance for large, flexible aircraft designs such as the YF-12. To remedy the problem, Lockheed engineers designed a dual-mode adaptive landing-gear system for the YF-12A. The configuration included a strut with an optimized air load-stroke curve during landing and an automatic switch-over system to allow for a flatter air load-stroke curve during taxi.

Lockheed engineers first proposed testing a dual-mode landing-gear system in 1976 and installed the prototype system in the YF-12A in January 1977. Pilots then flew three series of test runs, demonstrating three gear configurations. Don Mallick and Ray Young led off with two baseline landing-gear

The first NASA Blackbird crews included (from left to right) Ray Young, Fitz Fulton, Don Mallick, and Vic Horton. Young and Horton served as flight test engineers. (Courtesy of NASA)

stiffness tests in February. Fitz Fulton and Vic Horton then flew two tests with a large-volume auxiliary air chamber added to the system to modify the air load-stroke curve. Mallick and Young accomplished the final two tests to demonstrate a mixed-volume configuration. For each of the three configurations tested, the crew performed a series of eight taxi runs, providing constant-speed taxi data at a variety of airplane weights and velocities.

The study demonstrated the effectiveness of a dual-mode adaptive landing-gear system in reducing the dynamic response of an airplane during taxi. It also provided a database to aid in determining the degree of correlation between analytically predicted responses and actual test results with a full-scale YF-12A. During the tests, the system reduced dynamic response by 25% at the aircraft's center of gravity and as much as 45% at the cockpit. The research pilots who flew the aircraft commented that the "degree of ride improvement is quite noticeable, particularly at the higher gross weights."[129]

A final YF-12 landing study took place in March 1978. Dryden engineers scheduled three space shuttle orbiter landing approach simulation flights in the YF-12A. The first occurred on 23 March with Fulton and Horton as the crew. The last two flights took place on 31 March. Einar Enevoldson flew with Ray Young, and Bill Dana flew with Vic Horton. The pilots felt that the aircraft had exceptionally good control characteristics during shuttle-type approaches, which involved precise maneuvering just prior to touchdown. They did not experience any tendency toward overcontrol or pilot-induced oscillation (PIO).

Researchers compared data collected from the YF-12 flights with that accumulated by simulated shuttle approaches flown in the F-8 digital fly-by-wire (DFBW) aircraft and the NC-131H total in-flight simulator (TIFS). These simulations paved the way for approach and landing tests using the Space Shuttle *Enterprise*, which, in turn, furnished planners with data for the first space shuttle orbital flight test (OFT-1). Air-launched from a Boeing 747, Space Shuttle *Enterprise* completed five approach and landing tests (ALT) between August 12, and October 26, 1977. The final flight ended with a dramatic pilot-induced oscillation as the vehicle settled onto the concrete runway. In a presentation at what was by that time NASA Dryden on August 17, 1978, Milt Thompson summarized the results of the various shuttle landing simulation test programs: "The landing characteristics of the orbiter are conditionally acceptable for an OFT-1 Edwards recovery but unacceptable for general operational use."[130]

Under contract to NASA Dryden, Gary Teper, Richard DiMarco, and Irving Ashkenas of Systems Technology Incorporated (STI) compared data from the YF-12 shuttle simulations and the approach and landing tests. They found that the YF-12 pilots could correct the vehicle's flight path quickly and smoothly while maintaining desired altitude. By contrast, orbiter pilots in the Space Shuttle *Enterprise* experienced significant oscillations in attitude and altitude. The comparative analysis of the two vehicles identified such critical orbiter flight control characteristics as excessive time delay in the attitude response to pilot control inputs and degraded flight-path response to attitude changes associated with an unfavorable orbiter pilot location in the cockpit. The Systems Technology Incorporated study determined that moving the pilot location forward improved the pilot's ability to control the vehicle's sink rate and landing performance.[131]

FLYING LABORATORY

Because of the Blackbirds' unique capabilities, they were ideally suited to serving as flying laboratories, subjecting a wide range of experiments to conditions of high-altitude flight at Mach 3 speeds. NASA researchers used the two remaining aircraft to study boundary-layer flow effects, digital integrated controls, heat transfer, and drag effects. Other experiments included the evaluation of a maintenance monitoring and recording system, measurement of engine effluents for pollution studies, noise suppression tests, sonic boom effects, and testing of a series of structural wing panels designed by NASA Langley and fabricated by Lockheed.

Not every potential use of the YF-12 aircraft came to fruition. In 1971, NASA requested a Lockheed Advanced Development Projects (ADP) study on the feasibility of launching a hypersonic drone from the YF-12C. On 30 July 1971, Lockheed's Henry G. Combs and John R. McMaster submitted

In 1971, at the request of NASA researchers, Lockheed submitted a study on the feasibility of using the YF-12C to launch a hypersonic drone from a dorsal pylon. Lockheed's evaluation incorporated data derived from experience with the MD-21 project. (Courtesy of Lockheed Martin)

the results of the study, which was based on the design of NASA's proposed HT-4 subscale high-speed transport model. After investigating various drone locations (dorsal and ventral), mounting and separation methods, and launch and landing options, the two engineers offered their conclusions. The launch of a dorsally mounted drone could be realized but with certain provisions.

Combs and McMaster proposed a series of steps before attempting a drone launch. Predicted aerodynamic and drag data would have to be substantiated with an adequate wind-tunnel program, including the determination of an optimum drone launch angle with respect to the carrier aircraft. Certain modifications to the YF-12C structure, such as hard points for mounting the drone, were necessary. Modification of the fuel system would be required to maintain the center of gravity. Also, engineers had to determine operating restrictions to avoid exceeding the aircraft's structural load limits.

The proposed HT-4, a model of a high-speed transport design, was scaled down to a length of 600 inches. Powered by a Pratt and Whitney RL-10 engine, it was designed for launch at standard YF-12 cruise and altitude conditions.

Although the Lockheed Advanced Development Projects study declared the concept technically feasible, some individuals raised concerns. ADP Senior Vice President Kelly Johnson approved the report, but commented cautiously: "This is a progress report only. I want to talk to NASA about safety aspects before concrete action is taken to implement any construction."[132]

Johnson had reason for concern, based on Lockheed's experience with the MD-21 configuration. After the first D-21 launch in 1966, Johnson noted,

"Mainly, we demonstrated the launch technique, which is the most dangerous maneuver...[of] any airplane I have worked on."

The accident during the fourth launch attempt substantiated Johnson's statement. An investigation board concluded that launching a relatively large vehicle from the Blackbird at Mach 3 was extremely hazardous. In light of this tragedy, it is apparent why Johnson voiced concern over the YF-12C/HT-4 proposal. Ultimately, the HT-4 project never materialized, possibly as a result of Lockheed's experience with the D-21.[133]

But the Blackbirds did notable work as platforms for many other projects. In July 1973, following ground tests in NASA's High Temperature Loads Laboratory, the surviving YF-12A was instrumented for boundary-layer measurements along the lower fuselage. Engineers typically use a number of empirical theories to predict compressible turbulent-boundary-layer parameters. Because these theories produce substantially different values, they require additional data from flight and wind-tunnel experiments. NASA researchers fitted the YF-12A with boundary-layer rakes to collect such data during flight. Each rake consisted of an aerodynamic vertical pylon with hollow tubes arrayed on its leading edge. Each tube collected measurements at a different point within the boundary-layer region of airflow. Instruments also took static-pressure and skin surface measurements. To provide undisturbed airflow over the lower fuselage, all upstream protrusions and vents needed to be removed or faired over.

The YF-12A also carried an aft-facing step experiment to determine the drag penalty caused by aft-facing surface discontinuities in a thick boundary-layer region. Such discontinuities caused drag and shock-wave propagation at

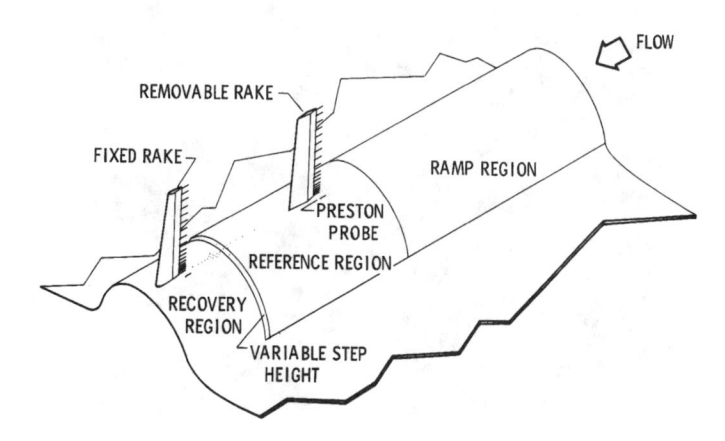

An aft-facing step experiment determined the drag penalty caused by surface discontinuities in a thick boundary-layer region. Vertical rakes contained probes to collect data throughout the boundary layer, at varying distances from the fuselage. (Courtesy of NASA)

supersonic speeds. The experiment provided designers with data for predicting the drag associated with lap joints and shingle structures on large aircraft at high Mach numbers. A panel on the aft fuselage created a small step over which the airflow passed. It consisted of a ramp region, a reference region, the step, a recovery region, and two boundary-layer rakes. Technicians installed pressure orifices along the plane of the surfaces of the reference and recovery regions and the step face. The step height—from 0.125 to 0.500 inches—could be varied between flights. The experiment produced data from incompressible speeds to ones in excess of Mach 3 and for ratios of boundary-layer momentum thickness to step height from 1 to 5. The results had applicability to the prediction of drag penalties for aft-facing discontinuities over a wide speed range for both the forward and aft regions of large airplanes.[134]

In early November 1974, the YF-12A underwent an experiment known as Coldwall to study the effects of compressible turbulent boundary-layer and heat-transfer coefficients at high speed. Designed and supported by NASA Langley, the apparatus consisted of a 13-foot-long stainless-steel tube mounted on a ventral pylon below the forward fuselage. The tube, equipped with thermocouples and pressure sensors, required cooling by liquid nitrogen and a covering made of insulating material. Planners intended the insulator to be pyrotechnically removed at Mach 3, exposing the tube to aerodynamic heating. Researchers also conducted wind-tunnel tests of a similar tube for comparison with data obtained in flight in order to validate ground research methods.

The Coldwall program suffered numerous setbacks. An initial low-speed functional check flight had to be aborted early when the experiment's accelerometer malfunctioned. A second attempt a week later ended the same way.

The YF-12A was fitted with an experimental pod called Coldwall to study heat transfer during Mach 3 flight. When the airplane reached cruise speed, the white insulating material was pyrotechnically removed from the pod to expose a nitrogen-filled stainless-steel tube to aerodynamic heating. (Courtesy of NASA)

The third flight, on February 27, 1975, took a more dramatic turn. On this day, the crew consisted of Don Mallick and Ray Young. Mallick climbed to 15,000 feet and advanced the throttle until the aircraft was traveling at 0.9 Mach number. At one point, the test card called for an aileron pulse with Roll/Yaw SAS off. As Mallick executed the maneuver, the ventral fin tore off, damaging the right wing and causing a fuel leak. The crew brought the YF-12 back to Edwards for a safe landing. An investigation determined that several factors contributed to the incident: 1) inadequate definition and documentation of aircraft limits; and 2) inadequate attention and adherence to existing, published, operating limits for the transonic region.

Clearly, because this test flight was to be conducted at known limit conditions and with a new configuration, planners should have taken a more conservative approach when scheduling such maneuvers as the aileron pulse that add considerable stress to the airframe.[135]

Test flights to demonstrate the structural characteristics of the Coldwall fixture resumed in July 1975. An operational check of the insulation removal system followed in August. Actual Coldwall data flights did not begin until August 1976, when a series of baseline "hot wall" flights occurred without the liquid-nitrogen coolant or insulation. The first true Coldwall flight took place on October 21, 1976. The premature loss of the insulation material prevented the collection of data on that flight, however. This problem continued to plague the experiment through June 2, 1977. The first good Coldwall data finally began to flow on June 23, when "the insulation removal system was successfully operated at the design test condition of Mach 3.0 and 72,500 feet."[136]

Without insulation, the super-cooled stainless-steel tube was instantly exposed to aerodynamic heating during high-speed flight. To validate predictive methods for estimating turbulent heat transfer, Engineers compared flight-data to calculations. (Courtesy of NASA)

The next Coldwall flight, on July 21, 1977, although less successful, was more dramatic. Insulation material from the experiment became ingested into the left engine of the YF-12A, resulting in an unstart. The right engine also unstarted. As the crew worked to restart the engines, the YF-12C chase aircraft also experienced multiple unstarts. Despite these problems, both aircraft returned safely to Edwards, but remained grounded until September for inspection and repair. The last two Coldwall flights occurred successfully on September 30 and October 13. At the time, several theories existed regarding the nature of turbulent heat transfer, but they yielded conflicting results when compared with wind-tunnel data. The competing schools of thought included Edward R. van Driest's equations for estimating turbulent heat transfer, E. R. G. Eckert's reference enthalpy method, and the Spalding-and-Chi method for determining skin-friction coefficients. Researchers used the Coldwall experiment data to validate Van Driest's theory.

Following the Coldwall flights, Don Mallick and Ray Young flew the YF-12A for a lower-surface boundary-layer survey and handling-qualities evaluation. Unfortunately, a malfunction of the wake visualization water spray system prevented them from obtaining any subsonic wake vortex flow data.

The incident of the lost ventral fin in February 1975 gave researchers an opportunity to flight test a new material. Technicians fitted a replacement ventral fin, made of Lockalloy skin panels over titanium framing, on the damaged YF-12A. Lockalloy, a metal alloy developed by Lockheed, consisted of 62% beryllium and 38% aluminum. Aircraft designers considered it

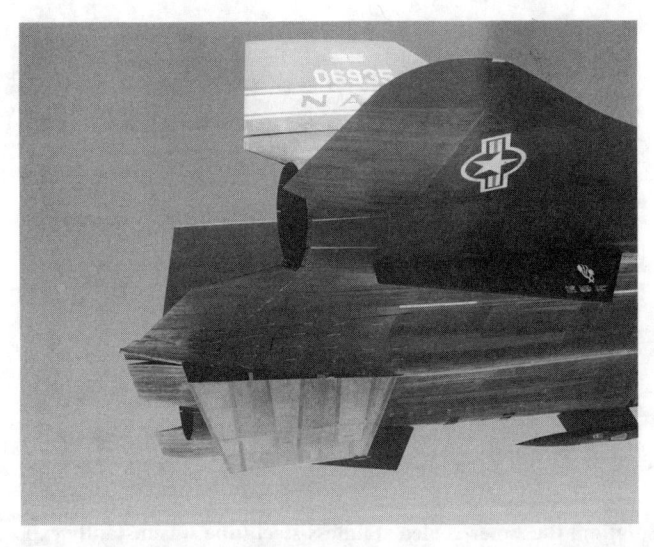

Loss of a titanium ventral fin in 1975 provided researchers with an opportunity to test a new material. In January 1976, the YF-12A flew with a ventral fin made from a Lockheed-designed beryllium–aluminum alloy called Lockalloy. (Courtesy of NASA)

a promising material for constructing high-temperature aircraft structures. The Lockalloy fin flew for the first time on January 16, 1976. Envelope expansion flights continued through May 20. Flight crews gradually exposed the new ventral fin to a variety of low-speed and high-speed flight conditions, eventually exceeding Mach 3. Loads and pressure distribution data closely agreed with predicted results. The Lockalloy skin increased the fin's torsional stiffness by 800% and its chordwise stiffness by 2000%. Its structural weight only increased by 38% (68 pounds).

The YF-12 also served as a testbed for advanced structural panels. A number of structural-configuration and material concepts showed promise in terms of reducing aircraft structural weight. Initially, a weldbrazed-skin stringer panel (one with stiffeners spot-welded in place) underwent approximately 37 hours of flight, including eight hours at Mach 3, with no adverse effects. Then, in July 1974, an experimental titanium honeycomb panel replaced an existing integrally stiffened primary structural part of the YF-12A upper wing surface, located over the landing gear. Researchers also arranged for flights with a composite panel of titanium honeycomb with a boron/aluminum face sheet onboard. These panel tests were initiated at NASA Langley under the Supersonic Cruise Aircraft Research (SCAR) program. Each panel type underwent ground testing by Lockheed before the actual flight test. In 1977, a SCAR panel was fitted to the YF-12C wing surface and flown regularly on a noninterference basis. Engineers assessed its structural integrity after each flight.[137]

WELDBRAZED TITANIUM SKIN STRINGER PANEL
PANEL WEIGHT = 3.17 KG (8.5 LB)

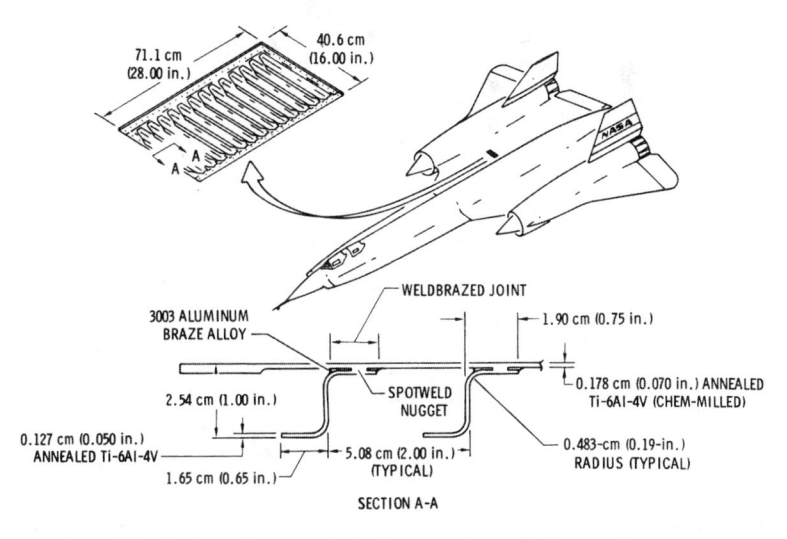

NASA used the YF-12 aircraft as testbeds for a variety of advanced structural panels of varying construction. (Courtesy of NASA)

In December 1977, several NASA research pilots who had not participated in the YF-12 program had the opportunity to fly the YF-12A. John Manke, Bill Dana, and Tom McMurtry flew familiarization flights with Vic Horton. Gary Krier and Einar Enevoldson made short hops with Ray Young. Both Dana and Enevoldson made additional flights as part of the space shuttle approach simulation program in March 1978. After that, the aircraft was grounded for installation of a shaker vane system.

As early as 1970, Lockheed proposed testing a loads alleviation and mode suppression (LAMS) system on a YF-12A and conducted a feasibility study. The design involved the use of small canards (or shaker vanes) on the aircraft forebody to excite the airplane's structural modes using controlled dynamic inputs at selected flight conditions. Such a system allowed a pilot to use feedback control techniques to suppress the aircraft's aeroelastic contributions to local acceleration and to develop techniques to reduce aircraft damage from air turbulence. The resulting flight-test data could then be compared with calculated aeroelastic response data and thus validate analytical techniques. The YF-12A design did not require LAMS technology itself and was, in fact, not well suited for evaluating rigid-body load alleviation techniques such as direct lift control. However, as a flexible aircraft it could be used to evaluate suppression techniques for aeroelastic modes. According to a study of LAMS for the YF-12, "evaluation of mode suppression techniques on the YF-12A

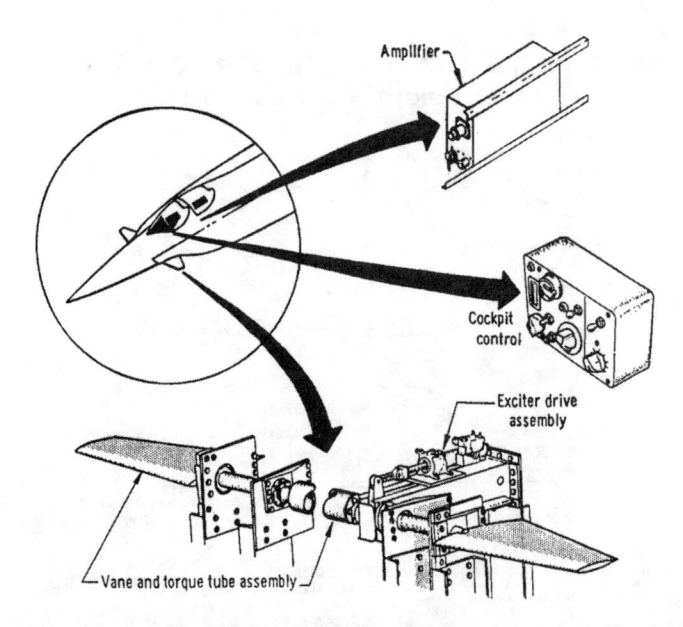

Small canards, called shaker vanes, were installed on the YF-12A as part of a loads alleviation and mode suppression system. Engineers compared structural mode data from research flights with computer-generated data to validate analytical models. (Courtesy of NASA)

could result in eliminating most of the design risk associated with LAMS technology and, thus, lead to acceptance by airframe manufacturers."[138]

The YF-12 mode suppression effort expanded LAMS research already begun with the NASA/Air Force XB-70 and the Air Force's NB-52E control configured vehicle (CCV) testbed. A Lockheed study in 1972 compared five candidate LAMS systems for the YF-12, evaluating each in terms of performance and mechanization. The Lockheed engineers concluded that a combination of small canards near the cockpit and inputs to the outboard elevons would provide the most effective LAMS system for the YF-12 aircraft.[139]

While the YF-12A remained in the hangar, the YF-12C returned to service. It had been unavailable since October 1977 in preparation for cooperative-controls (Co-Op) research. This program focused on digital integrated control of the aircraft's inlets, autopilot, autothrottle, air-data system, and navigation system.

Preparations for the cooperative-controls project began in early 1977, including tests of the aircraft's autothrottle system designed to control Mach and altitude simultaneously. In March 1977 Gene Matranga observed, "The system provided precision of control significantly better than that attainable by manual control at Mach 3.0 and constitutes a virtual technological breakthrough in flight path control for supersonic cruise aircraft." By June, he felt that "sufficient data had been obtained to verify the operational characteristics of the system, and to define control logic to be implemented in the upcoming Co-Op Digital Control Program." By September the NASA team had acquired all of the baseline propulsion/airframe interaction data necessary to prepare for the cooperative-controls study.

As a result, a cooperative airframe/propulsion control system (CAPCS) digital computer went into the YF-12C. This system incorporated the air-data, inlet, and autopilot systems into a single computer to improve overall aircraft flight control. Following installation of the CAPCS, Don Mallick and Ray Young took the aircraft up for a functional check flight on May 26, 1978. Gene Matranga later reported, "All flight objectives were satisfactorily accomplished with air data calculations, inlet control, and autopilot demonstrated." The flight also verified the basic logic flow and system architecture for the CAPCS.

Nine more CAPCS flights followed. Seven of them proved to be successful. The final CAPCS flight, on September 28, 1978, was aborted early because of an engine bypass door failure that precluded acquisition of cooperative-control data. It turned out to be NASA's last flight of the YF-12C. On October 27, the aircraft was returned to the Air Force after being ferried to the Lockheed facility in Palmdale. The YF-12C ended its NASA career on a high note. The CAPCS system exceeded designers' goals. Flight-path-control precision improved by a factor of 10. Additionally, aircraft range increased by seven percent, and inlet unstarts became almost unknown. Ultimately, Lockheed installed the system in the entire operational SR-71 fleet.[140]

To undertake research on the structural design required for a future supersonic transport aircraft, Jim McKay of the NASA Flight Research Center Dynamic Analysis Branch submitted a request for project approval on February 4, 1975. With a view toward future supersonic-transport development, NASA engineers needed accurate structural design calculations for flexible, low-aspect-ratio aircraft in the transonic flight regime. A large body of static and quasi-static test data had been generated during flight and ground YF-12 loads research. Additionally, McKay pointed out that the YF-12 team had "developed one of the most complete finite element structural analysis (NASTRAN) programs ever assembled for an aircraft, along with a complete static aeroelastic analysis program (FLEXSTAB)."[141]

Software engineers at NASA Goddard Space Flight Center in Greenbelt, Maryland, originally designed the NASA structural analysis (NASTRAN) program for space vehicles. In its first major application to an airplane, Alan Carter of NASA asked Lockheed's Al Curtis to create a NASTRAN model to support YF-12 loads research. The FLEXSTAB program, developed by Boeing for the supersonic transport, allowed researchers to assess the effect of airframe flexibility on stability and control characteristics of a supersonic aircraft. Perry Polentz of NASA Ames also sought out Curtis, on this occasion, to model the YF-12 using FLEXSTAB. Although Curtis encountered some problems adapting the program to the YF-12 wing configuration, the extensive analytical database set the stage for the proposed flight research effort. Jim McKay thought the resulting data would have "direct application to low-aspect-ratio vehicles with close dynamic coupling between major components such as fuselage and wing." He also believed that it would "provide data input to the Langley Research Center landing loads/response study in support of the work on actively controlled landing gear."[142]

With a view toward future supersonic transport development, NASA engineers used the YF-12A to explore structural design calculations for flexible, low-aspect-ratio aircraft operating in the supersonic regime. (Courtesy of NASA)

Finally, in February 1978, Dryden received funding from NASA Langley and the Air Force Flight Dynamics Laboratory to support a structural dynamics program with the remaining YF-12A. Installation of the canards began in April. The system consisted of a set of oscillating canards attached by a shaft to a driving unit in the nose of the airplane. The shaker vanes supplied controlled dynamic inputs to excite the response of various structural modes at selected flight conditions. On November 22, the YF-12A made its first flight with the new shaker vanes. Five more flights occurred by March 15, 1979. The shaker-vane study provided flight data on aeroelastic response, allowed comparison with calculated response data, and thereby validated analytical techniques. These last missions concluded NASA's first Blackbird flight research program.[143]

BACK IN BLACK

In 1990, after the Air Force formally retired the Blackbirds, NASA arranged to acquire two SR-71A models (Article 2022 and Article 2031) and SR-71B trainer (Article 2007) for use as research aircraft. At Dryden, Article 2031 was assigned NASA number 844. It arrived on February 15 and was followed by the second airplane, as NASA 832, on March 19. The aircraft were placed in storage while NASA crews gained flight experience in the SR-71B.

NASA research pilot Steve Ishmael and Lockheed's Rod Dyckman flew two functional check flights in the SR-71B (NASA 831) at Palmdale in early July 1991 before delivering the aircraft to Dryden on July 25.

In the early 1990s, NASA acquired two SR-71 aircraft, including the sole flyable trainer model. A third aircraft was returned to the Air Force for reactivation as an operational reconnaissance platform. (Courtesy of NASA)

Less than six months later, NASA 831 was used in tests of the Navy Radar Surveillance Technology Experimental Radar (RSTER) during a sortie over the Atlantic Ocean. The RSTER was a uhf sensor designed to detect high-flying missiles despite a high degree of radar clutter and jamming interference. In this, the only research mission for NASA 831, the NASA crew flew almost 6000 nautical miles in less than five hours.

In addition to Ishmael, Dryden flight crew members checked out in the Blackbirds included pilots Rogers Smith and Ed Schneider and flight-test engineers Robert Meyer and Marta Bohn-Meyer. Ishmael and Bohn-Meyer flew the first NASA mission in 844 on September 24, 1992. Marta Bohn-Meyer was the first and only female SR-71 crew member.

With the capability to fly at altitudes above 80,000 feet and at speeds in excess of Mach 3, the Blackbird was a unique research asset for Dryden. Additionally, the various fuselage bays that had held reconnaissance equipment were ideal for carrying research instrumentation and experiment packages. The first such items flown included an Orbital Sciences Corporation (OSC) frequency scanning experiment and the OSC F3SAT backup satellite, carried in passive mode to check prelaunch conditions.

At extremely high altitudes, the airplane was an ideal platform for remote-sensing technology experiments. Between October 1993 and October 1994,

NASA SR-71 crews included (from upper right to lower left) pilots Rogers Smith and Ed Schneider, as well as flight test engineers Bob Meyer and Marta Bohn-Meyer. Steve Ishmael, not shown, also piloted the aircraft. (Courtesy of NASA)

the aircraft carried several such packages. In the spring of 1993, the NASA 844 carried the Southwest Research Institute Ultraviolet Imaging System (SWUIS). The ultraviolet (UV) sensitive charge-coupled device, combined with a telescope, was a prototype for a miniature astronomical lab designed for use on the space shuttle. That summer the airplane's nose was equipped with a near-ultraviolet spectrometer (NUVS) for observation of volcanic gases in the UV spectrum. An upward-looking UV-sensitive video camera recorded a variety of celestial objects in wavelengths blocked from the view of ground-based astronomers. The Low Earth Orbit Experiment (LEOEX) validated technology for ozone mapping sensors to be carried on the Russian Meteor-3 satellite. Finally, a Dynamic Auroral Viewing Experiment (DAVE) provided data for the U.S. Navy.

The SR-71 also served as a testbed for an optical air-data system (OADS), a fiber-optic device using laser technology to replace the pitot tube (airspeed probe) on high-performance aircraft. It used laser light instead of air pressure to produce airspeed and attitude reference data such as angle of attack and sideslip normally obtained with small tubes and vanes extending into the airstream or from tubes with flush openings on an aircraft's outer skin. The flights provided information on the presence of atmospheric particles at altitudes above 80,000 feet, where future hypersonic aircraft might be expected to operate. The system, known as a sheet-pair laser anemometer, projected six sheets of laser light from the underside of the airplane. As microscopic atmospheric particles passed between the beams, direction and speed were measured and processed into standard speed and attitude references. An earlier optical air-data system was successfully tested at Dryden on an F-104 testbed.

The SR-71B served as a research platform, but was also used for crew training and proficiency. (Courtesy of NASA)

Under NASA's commercialization assistance program the SR-71 was used in the development of Motorola's commercial satellite-based, instant wireless personal communications network, called IRIDIUM. During IRIDIUM development tests, the SR-71 acted as a surrogate satellite for transmitters and receivers on the ground.

The SR-71 also was used in a project for researchers at the University of California–Los Angeles (UCLA), who were investigating the ability of charged chlorine atoms to protect and rebuild Earth's ozone layer.

Whereas NASA 844 was the research workhorse, the SR-71B was used primarily for crew training and proficiency. NASA 831 later set an unofficial milestone. During a checkout flight for pilot Ed Schneider on October 18, 1994, with Steve Ishmael as instructor pilot, the SR-71B reached an altitude of 84,700 feet. It was the highest-altitude NASA flight ever achieved in the Blackbird. The SR-71B continued to fly at Dryden until October 1997.[144]

Phoenix Rising

In 1994 U.S. Senator Robert Byrd, along with several members of the armed services and Congress, lobbied to reinstate several Blackbirds to operational service as a contingency reconnaissance capability. Two SR-71A airframes (Articles 2018 and 2022) were pressed into service while NASA and the Air Force agreed to share the SR-71B trainer. The new operational organization was designated Detachment 2 (DET 2) of the 9th Reconnaissance Wing. Headquartered at Beale Air Force Base near Marysville, California, DET 2 was stationed at Edwards to minimize operating costs and take advantage of the proximity of the NASA operation and the SR-71 flight simulator at Dryden.[145]

Article 2022 (briefly known as NASA 832) was never used for any NASA research missions. On January 12, 1995, Steve Ishmael and Marta Bohn-Meyer ferried the airplane to Lockheed's Palmdale facility. NASA crews subsequently flew Article 2022 at Palmdale in 1995 and 1996, conducting functional check flights in support of the SR-71 reactivation effort. For a mission June 14, 1996, Ed Schneider and Lt. Col Blair crewed Article 2022 during a four-hour sortie to test the Blackbird's advanced synthetic aperture radar, defensive avionics, and airborne data-link systems.

Also in 1995, NASA crews flew a number of sonic-boom research flights in support of the High-Speed Research Program (HSRP). The HSRP was a NASA-wide program to develop technology for a supersonic passenger aircraft called the high-speed civil transport (HSCT).

Researchers used the SR-71 to study ways of reducing sonic-boom overpressures in the hope that such data could eventually lead to aircraft designs that would reduce peak overpressures and minimize the startling effect they produce on the ground.

To measure the intensity of shock waves on aircraft in flight, NASA pilots flew an instrumented F-16XL in the wake of the SR-71 at supersonic speeds. (Courtesy of NASA)

Instruments at precise locations on the ground recorded the sonic booms as the aircraft passed overhead at known altitudes and speeds. For some flights an F-16XL aircraft was flown behind the SR-71, probing the near-field shock wave while instrumentation recorded pressures and other atmospheric parameters.[146]

While NASA continued with various research projects, three Air Force crews gained proficiency in the Blackbird. A third SR-71A (Article 2019) was readied for use, but the end was near for the Air Force program.

Opponents of the Blackbird reactivation exploited a complex loophole in the legislation, asserting it was technically illegal for the Air Force to operate the SR-71. Consequently the Pentagon suspended flight operations at DET 2 on April 16, 1996. This was overturned, however, as Blackbird supporters on the Senate Appropriations Committee threatened to defeat the Intelligence Authorization Act for fiscal year 1997, a move that would have halted all U.S. intelligence activities.

Subsequently $39 million was allocated to the Blackbird program, but opponents continued to lobby for cancellation. The end effectively came on October 15, 1997, when President Bill Clinton used line-item veto power to strike down the SR-71 funding. DET 2 personnel continued with routine maintenance work until September 30, 1997, the end of the fiscal year, and the detachment was officially deactivated on October 6, 1999.[147]

SWAN SONG

The final major project for NASA 844 involved the Linear Aerospike SR-71 Experiment (LASRE). Technicians mounted a 41-foot-long flight-test

For the Linear Aerospike SR-71 Experiment (LASRE), the aircraft carried a dorsally mounted scale half-model of the X-33, complete with a functional engine. The aerospike engine, however, was never hot-fired during the project. (Courtesy of NASA)

fixture, dubbed the "canoe" and capable of containing liquid rocket propellants, on top of the aircraft. It supported a 12% scale, half-span model of an X-33 research vehicle, complete with a working linear aerospike engine with eight thrust cells. In this way the SR-71 served as a flying wind tunnel.

The LASRE model was first flown on October 31, 1997, and NASA crews conducted several flights for aerodynamic and fuel flow data before the project was terminated in 1998. Although the X-33 project was cancelled, the experiment provided researchers with information that might help predict how operation of aerospike engines at altitude will affect the aerodynamics of a future reusable launch vehicle.[148]

In 1999, the modified aircraft was flown to evaluate its handling qualities with the test fixture in place and the model removed. The resulting data proved that the test fixture had little impact on aircraft performance. Unfortunately, no projects to employ the test fixture materialized, and the airplane's future looked grim.

NASA 844 made its final flight at the Edwards Air Force Base Open House and Air Show on October 9, 1999. Crewed by Rogers Smith and Robert Meyer, the aircraft attained a speed of Mach 3.21 and an altitude of 80,100 feet. Another flight was scheduled for the next day, but was cancelled because of a serious fuel leak. NASA 844 never flew again, nor did any other Blackbird.[149]

NASA became the last agency to operate the SR-71. The airplane made its last flight in October 1999. (Courtesy of NASA)

NASA 844 remained in flyable storage for one year pending possible future use, but the agency's Blackbirds were deemed too expensive to operate, or simply maintain in flyable condition, indefinitely. NASA officials ultimately concluded the SR-71 should be retired. It marked the end of an era. A Dryden ground crew placed NASA 844 on permanent static display at the Center on September 14, 2002.[150]

Chapter 6

LESSONS LEARNED

The Blackbirds provided numerous valuable lessons to designers, builders, and users of these remarkable aircraft. These lessons include results of the Lockheed Advanced Development Projects Division's "Skunk Works Principles" of management, innovative design and manufacturing techniques, and data from numerous experimental research programs.

KELLY'S WAY—THE SKUNK WORKS APPROACH

Since the early 1940s, the Lockheed (now Lockheed Martin) Skunk Works has become synonymous with innovative aerospace design and manufacturing techniques. This is largely the result of a management approach pioneered by Skunk Works founder Clarence "Kelly" Johnson. Designed to foster creativity and innovation, his method established principles for development and production of highly complex aircraft in a relatively short time and at relatively low cost. Although not easily applied in the corporate world of the early 21st century, it is worthwhile to study this innovative business model.

Johnson often summed up his method in just seven words: "Be quick. Be quiet. Be on time." Eventually, however, he wrote a set of 14 rules addressing program management, organization, contractor/customer relationships, documentation, customer reporting, specifications, engineering drawings, funding, cost control, subcontractor inspection, testing, security, and management compensation. These became the Basic Operating Rules of the Skunk Works:

> 1. The Skunk Works manager must be delegated practically complete control of his program in all aspects. He should report to a division president or higher. (It is essential that the program manager have authority to make decisions quickly regarding technical, finance, schedule, or operations matters.)

> 2. Strong but small project offices must be provided, both by the customer and contractor. (The customer program manager must have similar authority to that of the contractor.)

3. The number of people having any connection with the project must be restricted in an almost vicious manner. Use a small number of good people: 10 to 25 percent compared to the so-called normal systems. (Bureaucracy makes unnecessary work and must be controlled brutally.)

4. A very simple drawing and drawing release system with great flexibility for making changes must be provided. (This permits early work by manufacturing organizations, and schedule recovery if technical risks involve failures.)

5. There must be a minimum of reports required, but important work must be recorded thoroughly. (Responsible management does not require massive technical and information systems.)

6. There must be a monthly cost review covering not only what has been spent and committed, but also projected costs to the conclusion of the program. Don't have the books 90 days late and don't surprise the customer with sudden overruns. (Responsible management does require operation within the resources available.)

7. The contractor must be delegated and must assume more than normal responsibility to get good vendor bids for the subcontract on the project. Commercial bid procedures are very often better than military ones. (Essential freedom to use the best talent available and operate within the resources available.)

8. The inspection system as currently used by the Skunk Works, which has been approved by both the Air Force and Navy, meets the intent of existing military requirements and should be used on new projects. Push more basic inspection responsibility back to subcontractors and vendors. Don't duplicate so much inspection. (Even the commercial world recognizes that quality is in design and responsible operations—not inspection.)

9. The contractor must be delegated the authority to test his final product in flight. He can and must test it in the initial stages. If he doesn't, he rapidly loses his competency to design other vehicles. (Critical, if new technology and the attendant risks are to be rationally accommodated.)

10. The specification applying to the hardware must be agreed to in advance of contracting. The Skunk Works practice of having a specification section stating clearly which important military specification items will not knowingly be complied with and reasons therefore is highly recommended. (Standard specifications inhibit new technology and innovation, and are frequently obsolete.)

11. Funding a program must be timely so that the contractor doesn't have to keep running to the bank to support government projects. (Rational management requires knowledge of, and freedom to use, the resources originally committed.)

12. There must be mutual trust between the customer project organization and the contractor with very close cooperation and liaison on a day-to-day basis. This cuts down misunderstanding and correspondence to an absolute minimum. (The goals of the customer and producer should be the same—get the job done well.)

13. Access by outsiders to the project and its personnel must be strictly controlled by appropriate security measures. (This is a program manager's responsibility even if no program security demands are made—a cost avoidance measure.)

14. Because only a few people will be used in engineering and most other areas, ways must be provided to reward good performance by pay not based on the number of personnel supervised. (Responsible management must be rewarded, and responsible management does not permit the growth of bureaucracies.)

This management approach offers a proven, efficient method for developing new technologies, executing engineering and manufacturing development programs, procuring limited production systems at low rates, and upgrading current systems.[151]

Under these rules a project is organized around a manager who has total control of all aspects of the program. This provides the manager with the ability to control costs and meet rational milestones and objectives. Other functional organizations within the company such as human resources, information services, facilities, environmental health and safety, legal and other specialty areas provide on-demand support to the program manager.

The program's organizational structure is simple and contains built-in checks and balances. Overall staffing is kept to a minimum to provide clear lines of responsibility and maintain program security. The Skunk Works approach calls for the use of a small number of skilled personnel who are given broad responsibility and a substantial, but rational, workload. Keeping management and total staffing to a minimum results in greater individual responsibility and job satisfaction, improved communications, higher productivity, and reduced costs.

Success is dependent upon a cohesive team working closely together to achieve well-defined objectives. Tasks and responsibilities are clearly defined, and progress is measured and tracked using integrated plans and schedules. Managers of various subgroups must have a clear understanding of how their

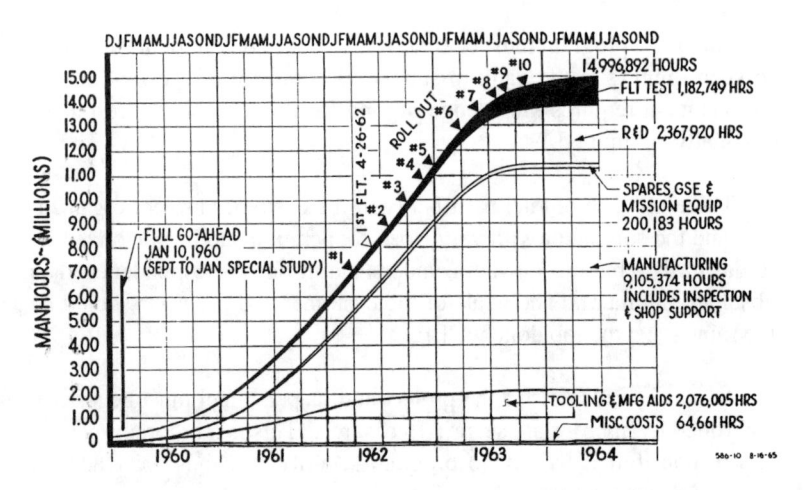

This chart illustrates the number of man hours expended on the A-12 program during the construction of 10 flyable and one static airframe. (Courtesy of Lockheed Martin)

role contributes to the success of the overall program. Formal weekly program reviews track the program's progress, whereas smaller meetings provide a forum for ironing out differences of opinion or improving operating procedures. If the program involves development of new or unique capabilities, participants must be willing to accept failures and incorporate changes based on lessons learned.

The Skunk Works approach works only if the customer is committed to working in a similar manner. This starts with a small, high-quality, highly responsive customer program office and a small supporting organization as needed. Like the contractor program manager, the customer program manager must also be given singular authority and broad responsibilities, reporting to a senior decision-capable management level. The contractor and customer teams should maintain open communications on program issues in order to foster teamwork, rapid joint problem solving, and mutual trust, rather than adversarial relationships. Small program offices and regular communications minimize the need for formal reports, documentation, and frequent program reviews.

If possible, contracts should be tailored to the specific procurements and eliminate restrictive and nonessential provisions while conforming to statutory and regulatory requirements. Increasing demands by government agencies, however, for contract provisions requiring extensive reporting, prior government approvals, and new administrative systems reduce a contractor's ability to tailor contracts in such a fashion.

Specifications should be as simple and brief as possible. Skunk Works practice emphasizes what is to be accomplished rather than how it is to be

CONTRACT & ENGINEERING RELATIONSHIPS

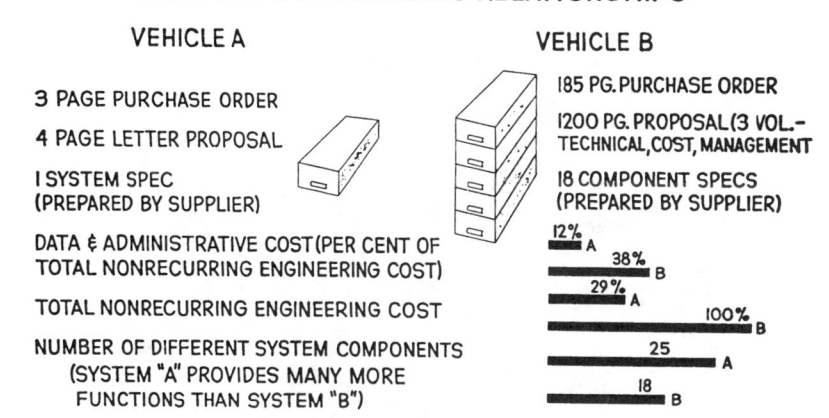

VEHICLE A
- 3 PAGE PURCHASE ORDER
- 4 PAGE LETTER PROPOSAL
- 1 SYSTEM SPEC (PREPARED BY SUPPLIER)
- DATA & ADMINISTRATIVE COST (PER CENT OF TOTAL NONRECURRING ENGINEERING COST)
- TOTAL NONRECURRING ENGINEERING COST
- NUMBER OF DIFFERENT SYSTEM COMPONENTS (SYSTEM "A" PROVIDES MANY MORE FUNCTIONS THAN SYSTEM "B")

VEHICLE B
- 185 PG. PURCHASE ORDER
- 1200 PG. PROPOSAL (3 VOL. - TECHNICAL, COST, MANAGEMENT)
- 18 COMPONENT SPECS (PREPARED BY SUPPLIER)

- 12% A
- 38% B
- 29% A
- 100% B
- 25 A
- 18 B

This graphic compares documentation requirements of a Skunk Works–type project (Program A) vs a project using conventional management methods. Reduction of paperwork is a hallmark of the Skunk Works management style. (Courtesy of Lockheed Martin)

accomplished, specifying only critical performance parameters as requirements. The original U-2 specification document was just 35 pages long. The SR-71 specification totaled 54 pages, relatively small and highly tailored compared to average procurement programs.

Manufacturing and quality-assurance personnel should be involved early, working closely with design, structures, and materials engineers to ensure product criteria are met. An integrated product development process ensures that the contractor will meet performance, quality, production, and affordability requirements. The engineering drawing system should readily accommodate change, and designers should be able to directly interface with manufacturers if changes are required. Modern computer-aided design techniques (not available when the A-12 was built) can reduce the need for full-scale mock-ups, and tooling should be kept to a minimum, especially for prototype programs. During fabrication and flight test, critical inspections verify compliance of processes with engineering requirements.

Flight testing of prototype and full-scale development (FSD) aircraft is conducted under the direction of a flight-test manager reporting directly to the program manager. Duties for this position include test planning, ground and flight testing, data acquisition and analysis, flight vehicle maintenance and support, and test-data documentation. The primary objective is to get test results as quickly as possible and apply lessons learned. The contractor is entirely responsible for testing prototype vehicles. During FSD testing, the work is accomplished by an integrated team of contractor and customer personnel.

Historically, Skunk Works programs have met very stringent security requirements. Development of the Blackbirds began as a covert special-access

Lockheed technicians work on final assembly of the YF-12A. (Courtesy of Lockheed Martin)

program. Only personnel with a strict need to know were briefed into the program. This policy not only protected national security but also prevented interference from outsiders and thereby increased productivity. A restrictive access policy can be similarly implemented on unclassified programs to improve efficiency and reduce costs.

By following these practices, the Skunk Works has consistently demonstrated the ability to design, develop, and produce highly advanced aircraft at low cost and in a minimal amount of time. The Skunk Works management principles allow a contractor to shorten the acquisition cycle and increase efficiency.[152]

Any organization can apply these principles, but of course applying the principles does not guarantee success. Historically, Skunk Works projects succeeded through a combination of factors including personal leadership, strong governmental support, organizational momentum, and a solid reputation based on a long history of achievement.

Kelly Johnson's strong leadership molded the Skunk Works into a lean, efficient organization. As a talented engineer, he was a technical leader as well as a manager. When engineering problems arose, Johnson took responsibility for addressing them personally. He motivated his staff through the strength of his personality, but had a legendary temper and a tendency to micromanage. He succeeded because he encouraged creativity and rewarded accomplishments.

In 1975, Ben Rich became head of the Skunk Works. A highly capable engineer in the field of thermodynamics and propulsion, Rich had been with the company since 1954 and had worked on the U-2, A-12, and SR-71. Because he lacked Johnson's breadth of technical expertise, he delegated more responsibility to his subordinates. He did, however, preserve Johnson's operating philosophy.

Good management is not the sole factor contributing to the company's success. The Skunk Works benefited from the strong support of its government customers. Because agencies such as the Central Intelligence Agency and Air Force were willing to abide by Johnson's basic operating rules, the company was able to successfully implement its streamlined contracting methods.[153]

Sometimes, the Skunk Works had to earn the customer's trust. By September 10, 1963, for example, the CIA's John McCone had become agitated by his perceived lack of progress in reaching OXCART design goals. He subsequently instituted an in-house study comparing Lockheed's rate of A-12 development with that of Convair's B-58. Results showed that the A-12 had been developed four times faster than the B-58 and had come closer to specified objectives in a given time, and at substantially lower cost.[154]

Over the years, the company developed a special relationship with its customers. By repeatedly demonstrating the ability to produce rapid, innovative results at low cost, the Skunk Works became the "go-to" organization for cutting-edge aerospace technology. Customers were able to bypass standard competitive procurement methods and easily justify sole-source contracts on the basis of the company's unique capabilities. A long history of innovation, in which successes far outweighed failures, gave customers the confidence to invest their money with the Skunk Works in development of high-risk, high-payoff technology.

In the years since Ben Rich's retirement, the Skunk Works has seen numerous organizational changes, especially since Lockheed merged with Martin Marietta in 1995. Adapting to corporate and legislative changes, Lockheed Martin's Advanced Development Programs (also known as The Skunk Works) Division continues its tradition of developing transformational strategies and classified products in a quick, quiet, and quality manner to support its varied customers. It continues to develop new innovations to serve our nation's defense for decades to come.

In the modern corporate structure, however, serious disconnects between acquisition policies and procedures have created an environment that is not usually conducive to true Skunk Works operations. The longer a program lasts, the harder it becomes to maintain a small organization. Attempting to exclude personnel who, for security reasons, do not have a need to know often leads to suspicion from outsiders. A tendency to trim near-term budgets by stretching program schedules undermines the program manager's ability to conclude the project as rapidly as possible. Teaming and subcontracting further challenge the program manager's ability to maintain a tight-knit organization and solve problems quickly. Although the Skunk Works approach

An Enormous Technological Achievement

The process of designing, building, and operating the unique family of aircraft known as the Blackbirds provided numerous technological lessons. Perhaps the most impressive characteristic of the Blackbirds is the fact that they were designed before the advent of supercomputing technology. A small team of talented engineers, using slide rules and know-how, built a family of operational airplanes capable of flying faster and higher than any airbreathing craft before or since. In addition, they had to invent new methods for parts fabrication, tooling assembly, construction, and testing.[156]

Because they needed to operate at ram-air temperatures of over 800°F, the Blackbirds were the first airplanes to be built primarily of titanium. Aluminum, commonly used for basic aircraft structure and skin panels, had too low a melting point. Stainless steel was excessively heavy. From a structural standpoint titanium was ideal: exceptionally strong, relatively light, and capable of retaining its structural integrity at extremely high temperatures. Unfortunately it was also costly, scarce, and difficult to work with. Initially, more than 80% of the titanium delivered to Lockheed was rejected because of metallurgical contamination, eventually traced to impurities in the local water supply. Lockheed technicians overcame these difficulties and pioneered new manufacturing techniques.[157]

Because of customer requirements for radar-cross-section reduction as well as for high speed, Lockheed spent a great deal of time and money investigating high-temperature radar-absorbing materials. These included pioneering

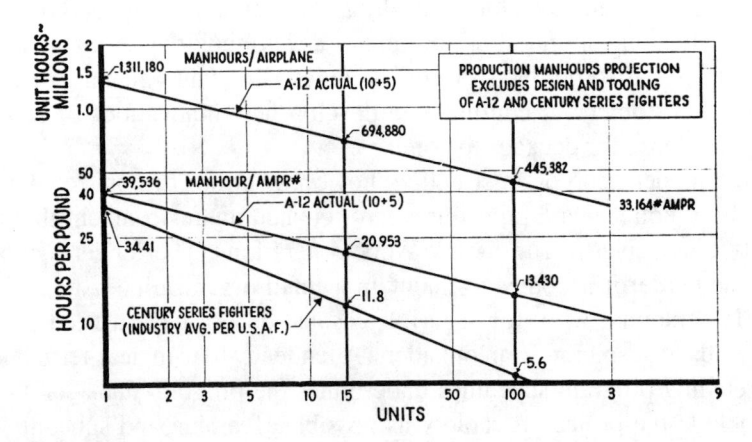

This chart compares the number of man hours spent on the A-12 vs those spent on contemporary fighter aircraft. (Courtesy of Lockheed Martin)

work with first-generation composites and high-temperature plastics. Large composite assemblies on the Blackbirds included the chines, tails, and inlet spikes, marking the first time such materials had been used as a major part of an aircraft's structure and the first use of such structures in a high-temperature environment. Lockheed's innovative methods of reducing total and incidental radar cross section became the basis for virtually all U.S. low-observables studies and hardware to follow, eventually leading to development of true "stealth" aircraft that would be virtually invisible to radar.[158]

The Blackbirds' hostile operating environment necessitated the development of fuels, lubricants, and sealants that could withstand high temperatures. At cruising speeds the airplane's fuel had to be stable at temperatures exceeding 350°F when it was fed from the tanks into the engine's fuel system. In subsonic cruise flight during aerial refueling, fuel temperature could drop as low as –90°F. Lubricating oil had to be effective at temperatures as high as 600°F while remaining suitably viscous at temperatures below 40°F.[159]

Other specialized materials, such as seals for the hydraulic system and high-temperature window glass, were developed for the Blackbird. For operational missions, it was necessary to solve the problem of how to take clear pictures through hot, turbulent airflow across the camera windows. It also took three years and $2 million to develop glass windscreen panes for the cockpit that would retain optical qualities at cruising speeds. This necessitated pioneering a unique process for metal-to-glass fusion using high-frequency sound waves.[160]

In building the Blackbirds, Lockheed designers had to take into account the fact that at cruise conditions high temperature gradients existed across large flat titanium skin panels and their attached substructure. Aerodynamic

The third A-12, Article 123, under construction at Burbank. (Courtesy of Lockheed Martin)

heating caused the metal panels to expand and buckle if attached to the structure using conventional methods. On the Blackbird's wing, the problem was solved using a standoff clip that lay between the wing skin and substructure. The clip retained structural integrity while providing a heat shield between the parts. Slip joints and special fasteners had to be provided to prevent high stress levels at the attachments points. Beaded and corrugated panels were used to minimize skin buckling.[161]

Because the A-12 went from limited go-ahead to first flight in just 30 months, the manufacturing process required a total overlap in all design and construction phases. Consequently, there was not enough time to do things progressively and in sequence. Design actions had to be taken based on the best-available estimates.

Manufacturing tasks had to be sequenced and completed on a least-risk basis in cooperation with engineering personnel. The engineers had to issue a "Basis for Structural Design" document as early as possible, with enough loading information to allow designers to set design details and release drawings to the manufacturing shop without delay. All components had to be designed as light as the Basis for Structural Design would permit. The document was updated whenever significant information came to light, and components that already had been built were changed as required.

Extensive component testing was accomplished as early as possible to validate the overall design as the airplane was being built. If testing dictated a change, it was accomplished as a progressive modification before the first flight.

Static test articles were subjected to critical loading conditions of at least 155% of design limit loads. Strain-gauge data validated or altered design requirements of individual vehicle components and structures. Static testing to ultimate design loads was conducted before performing limit load testing on the flight vehicle. Because most critical loading of the Blackbird's structure took place in the low-temperature transonic region, room-temperature testing on the static test article was adequate. In some cases, room-temperature overload was used to simulate hot conditions. In other cases, components were actually heated to design temperatures.

Lack of advanced computer modeling capability necessitated reducing problems to their lowest common denominator before making calculations. This simplified the airplane's design and forced engineers to rely on their expertise. According to Lockheed engineer John Alitzer, "The most valuable design engineer is the engineer with the ability to ballpark a design based on raw data and then have it prove to be close to correct when all supporting information is fed to the computers."[162]

Creating a powerplant for the Blackbirds also resulted in significant lessons for engine manufacturer Pratt and Whitney as the company embarked on an intensive design and development effort. The propulsion integration phase

Construction of the Blackbirds presented numerous technical challenges. (Courtesy of Lockheed Martin)

involved determining aerodynamic compatibility, installation and structural technology advances, and development of a unique mechanical power-drive mechanism and tailored fuel system. Engine designers had to explore uncharted territory and discover, identify, and address numerous challenges.

One of the greatest of these involved the issue of engine cooling. Engine inlet air temperature exceeded 800° F under certain conditions, and the fuel inlet reached 350°F. The temperature at the main and afterburner fuel nozzles periodically ranged from 600 to 700°F. Lubricants reached 700 to 1000°F in some localized parts of the engines. Because of the extremely hostile temperature environment, electronic components were limited to a fuel-cooled solenoid and a trim motor buried inside the fuel control mechanism. Consequently, control adjustments that were normally automated had to be made manually.

Another significant problem involved engine airflow. It quickly became apparent that a straight turbojet cycle provided a poor match for the inlet and did not produce the required net thrust at cruise Mach conditions. To overcome these difficulties, Pratt and Whitney designers invented a bleed-air bypass cycle to match engine/inlet airflow requirements. At speeds above Mach 2.0, the corrected airflow could be held constant at a given Mach number regardless of throttle position. Additionally, the bleed bypass cycle provided more than 20% additional thrust at high-Mach cruise.

Engine materials and fabrication technology presented some of the greatest challenges. The manufacturer had to learn how to form sheet metal from materials previously only used to forge turbine blades and devise methods for welding it successfully. Turbine disks, shafts, and other components also had to be fabricated from high-strength, temperature-resistant materials. Accessory drives, pumps, and other auxiliary equipment had to be designed to withstand the temperatures and stresses encountered in routine operation. Parts of the afterburner had to withstand as much as 3200°F.

Pratt and Whitney teamed up with the Ashland, Shell, and Monsanto companies to develop special fuel and chemical lubricants. The fuel itself served as the engine's hydraulic fluid, but there was nothing to cool the fuel; it made just one pass through the hydraulic system before being burned in the engine. Because the fuel was not lubricious, a small amount of fluorocarbon was added, allowing the fluid to properly lubricate various pumps and servos.

Developmental testing posed problems, as well. None of the available test facilities could provide steady-state temperature and pressure conditions necessary for testing the engine at maximum operating conditions. Pratt and Whitney technicians found a partial solution by routing the exhaust of a J75 engine through and around a J58 to simulate temperature transients. It was also necessary for them to develop instrumentation rugged enough to obtain accurate real-time measurements and better calibration facilities to process data.

Although Pratt and Whitney had a very large computer system for its day, it was no more sophisticated than some of the handheld calculators that became available within two decades. Consequently, like the Blackbirds it powered, the J58 was essentially designed by slide rule.

Once the engine was integrated with the airframe, ground and flight testing provided numerous lessons. Problems with engine start resulted in the addition of suck-in doors and engine bleed vents on the nacelles. In flight the engine was found to burn too much fuel at transonic speeds. Analysis of the problem revealed that the exhaust ejector section was the first part of the aircraft to exceed Mach 1.0, resulting from a design based on flawed wind-tunnel data. A pilot inadvertently solved the problem by attempting transonic speeds at a lower altitude and higher knots equivalent airspeed. The resulting lesson was to not run wind-tunnel tests of the nacelle unless the model included a simulation of adjacent aircraft surfaces and to take an adequate number of transonic data points.

At higher Mach numbers, new problems arose. Gearbox mounts started to exhibit heavy wear and cracking, and the driveshaft began to show twisting and spline wear. By placing simple styluses and scratch plates, technicians were astonished to discover the gearbox moved about four inches relative to the engine, much more than the shaft could tolerate. Addition of a double universal joint solved the problem. When the aircraft's fuel system plumbing immediately ahead of the engine started to show fatigue and distortion, measurements with a fast recorder indicated pressure spikes at the engine fuel inlet were off scale because of feedback from the engine hydraulic system. This problem had not manifested during static tests or aircraft ground runs

Extensive flight testing was necessary to discover and solve design problems before the Blackbird was declared operational. (Courtesy of Lockheed Martin)

because of a disparity in fluid volumes. Lockheed solved this problem by inventing a "high-temperature sponge" for installation in the accumulator ahead of the engine, which reduced pressure spikes to tolerable levels.

Unstart problems plagued the Blackbirds until several causes were identified. These included use of manual engine trim, nacelle leakage, and *g*-loading effects on angle-of-attack data from the nose probe. A number of changes eventually eliminated the problem. These included improved inlet and bypass door seals, addition of an engine auto-trim system, addition of an unstart signal to the derichment valve to protect the turbine, increasing the area of the inlet bypass doors that routed air to the ejector, and adding a *g*-bias to the inlet control and automated inlet restart procedure for both inlets regardless of which unstarted. Research in these areas also resulted in methods for reducing drag through improved nacelle sealing and bypassing air over the engine to the aft bypass door during normal flight.

Design problems were sometimes overlooked until after the hardware was built and flight tested. Pratt and Whitney designers became so obsessed with the problems of fuel heating that they neglected to take into account the fact that cold fuel was sometimes introduced into a hot environment and vice versa. This caused difficulties for the engine fuel control mechanism. To correct this, the main engine control system had to be insulated and redesigned so that it would respond only to fuel temperature.

Both Lockheed and Pratt and Whitney spent many hours integrating the inlet/nacelle structure and engine so that blow-in doors, bleed ducts, air-conditioner turbine drive discharge, etc., would not adversely affect any of the engine control sensors inside the inlet. To save time in the design process, Lockheed built the inlets as mirror images of one another, but the engines were made to be interchangeable (rather than dedicated left and right engines). The air-conditioner turbine discharge was located 45 degrees from one side of the engine centerline, and the engine temperature bulb was located at a 45-degree angle on the opposite side, placing the 1200°F turbine discharge in different locations relative to each nacelle. As a result, one engine always ran faster than the other until the mystery was solved and the problem fixed.

Pratt and Whitney's success was primarily the result of compatible conceptual designs, diligent application of engineering fundamentals, freedom to change the engine and aircraft design with minimal contractual paperwork, and exceptional teamwork. The customer gave both aircraft and engine contractors the freedom to take any actions they deemed necessary to solve problems as they arose. Government participation in the engine development was limited to a small cadre of highly qualified personnel who were oriented toward understanding the problems and approaches to solutions, rather than substituting their judgment for that of the contractors. Requirements for government approval as a prerequisite to action were limited to those changes involving significant cost or operational impact. This allowed the contractors to respond quickly to problems and develop innovative solutions without interference. By reducing the time between formal release of engineering specifications and production of hardware, the program's progress was accelerated and costs reduced.[163]

Matching the powerplant to the airframe was a significant challenge that centered around development of the inlet system. The Blackbird's inlet was a triumph of engineering that required extensive development work prior to finalizing the design. To arrive at a usable inlet configuration, Lockheed technicians collected approximately two million data points in the wind tunnel. At least that much data were later collected during flight testing for the purpose of validating predictive methods.[164]

Because the customer recognized that many problems involving the engine, airframe, or both could best be solved by a joint engineering effort involving both contractors, contracts were written to allow this type of activity without penalties. This method of operation allowed problems to be solved promptly. A more cumbersome management system might have resulted in costly delays or forced inappropriate compromises because of contractual interpretations. In the case of the Blackbirds, the management system resulted in shorter development time, rapid response to problems in the field, reduced retrofit costs, and earlier availability of production systems incorporating corrections for problems discovered during testing.[165]

RESULTS OF THE NASA BLACKBIRD RESEARCH PROGRAMS

Use of the Blackbirds as flying laboratories provided important lessons for researchers. The joint NASA/Air Force YF-12 research program of the

NASA used the Blackbirds as flying laboratories to validate predictive methods and explore high-performance flight characteristics. (Courtesy of NASA)

1970s and later SR-71 projects produced a wealth of data, derived from flight and ground research, as well as from simulation and modeling. Collectively, these investigations made important contributions to the advancement of aerodynamics and thermodynamics. Among other achievements, the comparison of flight data to wind-tunnel data and predictions helped researchers develop more accurate modeling techniques for flexible, supersonic aircraft designs.

At the start of the undertaking, NASA engineers and technicians faced a difficulty inherent in this type of research. Because the high speeds of the YF-12 generated sustained aerodynamic heating, the YF-12 team needed to devise data recording techniques suited to these conditions. Indeed, NASA project manager Gene Matranga noted that instrumentation for the YF-12 was "very inadequate to begin with, and it cost a lot of money to develop new instrumentation and make it work."[166] Once this problem was resolved, a wide vista of research opportunities became available.

To begin, researchers compared heating measurements in the High Temperature Loads Laboratory to in-flight heating. The results allowed them to calibrate more accurately instrumentation for loads measurement on high-speed aircraft by separating thermal loads from aerodynamic loads. They also explored structural configurations relative to the thermal environment, studied the resulting thermal stresses, and demonstrated how a thermal calibration of the aircraft eliminated heating effects from loads equations. Additionally, Matranga and his colleagues discovered unexpected hot spots and leakage of hot air into the aircraft's internal structure.[166]

The YF-12 aircraft likewise provided a wide range of propulsion data on variable-cycle engine operation and mixed-compression inlet operation. The flight research demonstrated that an inlet could be designed using small-scale models and also showed that YF-12 inlet dynamics had a profound effect on stability and control. The dynamics sometimes surprised the researchers. Air from the forward bypass doors, for instance, actually moved forward at least 12 inches before mixing with the boundary layer. "This startled everybody," said Matranga. "We didn't realize there was so much separation just ahead of the bypass area." Ultimately, NASA developed a computer control system for the bypass doors to improve efficiency. It increased aircraft range and performance and eventually became incorporated into the operational SR-71 fleet.[167]

Flight-test data also indicated that, during Mach 3 flight, air venting through the inlet bypass doors accounted for half of the aircraft's total drag. Inlet control proved to be imprecise because of sensor placement and accuracy, making unstarts a frequent occurrence. But NASA's YF-12 research program virtually eliminated unstarts through a combination of inlet spike schedule refinements and hardware improvements. This research led to development of a digital automatic flight and inlet control system (DAFICS). Consisting of four digital-computer-controlled subsystems and a computer-analyzer subsystem, DAFICS was eventually incorporated into the entire operational SR-71 fleet in the early 1980s.[168]

Wind-tunnel model data provided an opportunity to validate scale and wind-tunnel effects against the flight data and also enabled engineers to determine more precisely the placement of instruments in the airplane inlet. Inlet-flow-systems interaction studies helped researchers define the inlet operating envelope and yielded information about unstart/restart boundaries. Engineers compared data from the NASA Ames and NASA Lewis wind tunnels to data obtained during research flights to better evaluate scaling and tunnel effects. NASA researchers found that testing limitations prevented the attainment of identical test conditions in the mixed-compression inlet mode. To solve this problem, James D. Brownlow, Henry H. Arnaiz, and James A. Albers developed a mathematical model from which comparisons could be made using statistical techniques.[169]

During a series of landing studies, a mixed-volume dual-mode gear system reduced airplane dynamic response during high-speed taxi. The test procedure called for taxiing at constant speed on the same section of runway during each run. In practice, however, the airplane traversed slightly different sections of runway from one test to the next. Aircraft weight differed throughout the test series as a result of fuel consumption. Although the mix of aircraft weights and velocities used did not represent any one specific aircraft configuration, the results encompassed a wide range of operational conditions. Overall, the dual-mode system provided significant dynamic response reductions, yielding a smoother ride. Analytical results generated by a digital computer program provided excellent correlation with the flight-test data at most areas, except the cockpit.

During handling-qualities investigations made during the YF-12 program, researchers concentrated on characteristics associated with longitudinal control during high-speed supersonic cruise, with possible application to the development of a supersonic passenger transport aircraft. Part of the investigation involved altitude hold and Mach hold, important because maximum range performance depended in part on accurate control of altitude and Mach number. In flight tests with a modified altitude and autothrottle hold mode, the YF-12C proved to be the most stable aircraft platform thus far demonstrated at Mach 3 speeds.

Flight research on the YF-12 also included certain propulsion and aerodynamics problems encountered during the Blackbird's operational life. During flight at a constant power setting, for instance, many aircraft experienced a natural oscillation, called a phugoid. NASA research pilot Milton O. Thompson noted that with the YF-12 in particular,

> Automatic engine inlet operation can have a pronounced effect on phugoid characteristics. In fact, under certain conditions, the automatic inlet operation can couple with the autopilot system in a manner to drive the phugoid unstable. Another unanticipated problem is the effect of engine bypass air on stability and control characteristics. At certain flight conditions, engine bypass air being ducted overboard can produce rolling moments comparable to that available with full aileron deflection. Inlet unstarts are not uncommon even under so-called ideal steady state cruise conditions. Atmospheric anomalies may be responsible for some of these inlet unstarts, since we have seen some rather abrupt changes in outside air temperature at these high speeds; however, other unstarts are not readily explained.[170]

The YF-12 team also employed the aircraft as a platform to study human factors in a high-altitude supersonic cruise environment. Researchers first identified sets of aircraft and physiological parameters most sensitive to pilot workload. Next, they isolated and quantified physical and nonphysical workload effects. Finally, they gathered and reduced flight data for comparison with findings from a clinical study to develop a pilot workload model from which predictions could be made.

NASA engineers conducted extensive studies of the boundary layer, using instrumented rakes. They found significant discrepancies between wind-tunnel model data and flight-test results. These apparently resulted from surface imperfections on the full-scale aircraft. "There were lumps, and bumps, and waves," said Matranga. "A wind tunnel model was a nice, smooth, rigid structure." Although researchers developed predictive methods to compensate for

Because of its ability to cruise at Mach 3 speeds and altitudes above 70,000 feet, the Blackbird was a unique asset to NASA. (Courtesy of NASA)

these differences, according to Matranga "it requires interpretation on the part of very skilled people."[171]

Another benefit of the YF-12 program arose from the extreme altitude range at which the aircraft flew. NASA engineers Terry Larson and Jack Ehrenberger—working in support of the SR-71 Category II tests—extensively documented upper-atmosphere physics. Such atmospheric modeling proved useful for later designs of high-altitude research aircraft. The YF-12 contributed valuable sonic-boom information and showed that mild turbulence could be encountered even at the highest altitudes at which the aircraft cruised.[172]

Finally, NASA and Air Force researchers gained valuable information concerning loads suppression and mode alleviation for flexible aircraft using a nose-mounted canard (shaker vane) system. This apparatus allowed them to make direct comparisons with calculated aeroelastic response data and thereby validate available analysis techniques. Once the shaker vane study had been completed, the NASA YF-12 research program essentially ended.

Thus, over its 10-year life span, the program made significant contributions to high-speed aeronautical research. Perhaps most important, it left a legacy of structural, aerodynamics, propulsion, and atmospheric physics data likely to serve as the basis for future high-speed aircraft designs and analytical model evaluation.[173]

In the 1990s, NASA again employed Blackbirds in a variety of projects. With the ability to fly at altitudes above 80,000 feet and at speeds in excess of Mach 3, the Blackbird was a unique asset for high-speed, high-altitude

Here, a NASA SR-71 deploys its 40-foot drag chute during landing. (Courtesy of NASA)

research. Additionally, the various fuselage bays designed to hold reconnaissance equipment were ideal for carrying research instrumentation and experiment packages, some supporting the space program. One such item included a small satellite, carried to prelaunch conditions before actually being placed onboard a rocket.

The SR-71 also served as a platform for a miniature astronomical laboratory designed for use on the space shuttle. Additionally, the airplane provided a testbed for an optical air-data system (OADS). This device, using laser technology, would replace the conventional pitot tube (airspeed probe). Such technology has applications to vehicle designs requiring elimination of external protruberences, such as high-performance aircraft or aerospacecraft.

Because of its ability to fly at extremely high altitudes, the SR-71 was an ideal platform for remote-sensing technology experiments. Several such packages included experiments for observation of volcanic gases in the UV spectrum, ozone mapping sensors to be carried on a Russian satellite, and an optical sensor for viewing the Aurora Borealis.

Because of its speed capabilities, the SR-71 was useful in a series of sonic-boom research flights to develop technology for a supersonic passenger aircraft.

The most ambitious project, however, involved using the Blackbird as a flying wind tunnel. For the Linear Aerospike SR-71 Experiment, technicians mounted a 41-foot-long flight-test fixture on top of the aircraft. Capable of containing liquid rocket propellants, it supported a scale half-span model of an X-33 research vehicle, complete with a working linear aerospike engine. Although the engine was never fired, several research flights provided valuable aerodynamic and fuel flow data for engineers developing the X-33.

In its final research program, the NASA SR-71 carried a dorsal flight-test fixture. Here, it makes a pass over the Lockheed Martin facility in Palmdale. (Courtesy of NASA)

Finally, the modified aircraft was flown to evaluate its handling qualities with the test fixture in place and the model removed. The resulting data proved that the test fixture had little impact on aircraft performance.[174]

Shortly after completing this research program, the Blackbird was retired for the last time. Its legacy, however, will live on in future designs of high-performance aircraft and advanced aerospace vehicles.

Appendix A

BLACKBIRD PRODUCTION SUMMARY

This section provides a brief production history of all Blackbird variants, including the D-21 drone. A total of 50 manned aircraft and 38 drones were built.

Table A1 A-12 (Total built: 13)

Model	Serial no.	Article no.	History
A-12	60-6924	121	First flight: 26 April 1962. Initially flew with J75 engines. Testbed for airworthiness and handling qualities, envelope expansion, airframe/powerplant integration, subsystems, and propulsion. Served as resident test article, bailed to contractor. 322 flights (418.2 hours). Stored at Air Force Plant 42, Palmdale, CA, 6 June 1968. Placed on permanent display at Blackbird Airpark, Palmdale.
A-12 (A-12B)	60-6925	122	Used for radar-cross-section pole testing from June to November 1962. First flight: 15 January 1963. Initially flew with J75 engines. Served as controls testbed. Modified to SR-71 standards between April 1966 and April 1967, and redesignated A-12B. 162 flights (127.9 hours). Stored at Palmdale 16 September 1967. Displayed at U.S.S. Intrepid museum, New York, NY.
A-12	60-6926	123	First flight: 9 October 1962. Flew only with J75 engines. Used for project pilot proficiency and systems tests. 79 flights (135.3 hours). Crashed 24 May 1963.
A-12T (TA-12)	60-6927	124	Two-seat trainer. Designated A-12T by Lockheed, but listed as TA-12 in pilots' official flight records (Air Force Form 5). First flight: 22 January 1963. Flew only with J75 engines. Used for pilot checkout, instructor training and project pilot proficiency. 614 flights (1076.4 hours). Stored at Palmdale 29 May 1968. Displayed at California Museum of Science, Los Angeles, CA.

(*Continued*)

Table A1 A-12 (Total built: 13) (*continued*)

Model	Serial no.	Article no.	History
A-12	60-6928	125	First flight: January 1963. Initially flew with J75 engines. Used for project pilot proficiency and tactical practice. Flew fastest known A-12 flight: Mach 3.29 (2171 mph) on 8 May 1965. 204 flights (334.9 hours). Crashed 5 January 1967.
A-12	60-6929	126	First flight: March 1963. First A-12 initially equipped with J58 engines. 105 flights (169.2 hours). Crashed 28 December 1965.
A-12	60-6930	127	First flight: March 1963. Prepared for Operation SKYLARK (planned Cuban overflights) in 1965. Deployed to Okinawa, Japan, for Operation BLACK SHIELD overflights of Southeast Asia from 24 May 1967 to 9 June 1968. Completed 17 operational missions. Stored at Palmdale 12 June 1968. 261 flights (499.2 hours). Displayed at Alabama Space and Rocket Center, Huntsville, AL.
A-12	60-6931	128	First flight: June 1963. Prepared for Operation SKYLARK in 1965. Participated in SCOTCH MIST wet weather testing at McCoy AFB, FL, in 1965. Stored at Palmdale 1 July 1968. 232 flights (453.0 hours). Displayed at Minnesota Air Guard Museum, St. Paul, MN, in 1991. Transferred to CIA Headquarters, Langley, VA, in 2007.
A-12	60-6932	129	First flight: September 1963. Initially used for envelope expansion testing. Participated in SILVER JAVELIN sustained cruise (1.25 hours at or above Mach 3.1) mission on 27 January 1965. Set unofficial altitude record of 90,000 feet on 14 August 1965. Deployed to Okinawa for Operation BLACK SHIELD overflights of Southeast Asia 26 May 1967. Completed 7 operational missions. 268 flights (410.0 hours). Lost in the South China Sea on 5 June 1968.
A-12	60-6933	130	First flight: December 1963. Used for project pilot tactical practice. 217 flights (406.3 hours). Stored at Palmdale 5 June 1968. Displayed at San Diego Aerospace Museum, San Diego, CA.
A-12	60-6937	131	First flight: 19 February 1964. Used to test electronic countermeasures and side-looking radar. First A-12 deployed for Operation BLACK SHIELD from 23 May 1967 to 14 June 1968. Completed 20 operational missions. Last A-12 to fly. 183 flights (351.1 hours). Stored at Palmdale 21 June 1968. Displayed at Southern Museum of Flight, Birmingham, AL.

(*Continued*)

APPENDIX A

Table A1 A-12 (Total built: 13) (*continued*)

Model	Serial no.	Article no.	History
A-12	60-6938	132	First flight 4 March 1964. Prepared for Operation SKYLARK in 1965. 197 flights (370.0 hours). Stored at Palmdale 1 June 1968. Displayed in Mobile, AL.
A-12	60-6939	133	First flight 18 March 1964. Served as test aircraft. Never flew above Mach 3. 10 flights (8.3 hours). Crashed on landing approach 9 July 1964.

Table A2 YF-12A (Total built: 3)

Model	Serial no.	Article no.	History
YF-12A	60-6934	1001	First flight 7 August 1963. 180.9 flight hours Damaged beyond repair on 14 August 1966 during landing accident. Rear half was later used to build the SR-71C. Front half scrapped.
YF-12A	60-6935	1002	First flight 23 November 1963. Participated in joint USAF/NASA research program 1970 to 1979. 534.7 flight hours Displayed at National Museum of the U.S. Air Force, Wright–Patterson AFB, Ohio.
YF-12A	60-6936	1003	First flight 13 March 1964. Set official speed record of Mach 3.14 (2,070 mph) and altitude record of 80,257 feet on 1 May 1965. Participated in joint USAF/NASA research program 1969–1971. 439.8 flight hours Crashed 24 June 1971.

Table A3 M-21 (Total built: 2)

Model	Serial no.	Article no.	History
M-21	60-6940	134	First flight 22 December 1964, carrying D-21 (Article 501). Used only for captive-carry flights and chase for last flight of Article 135. 94 flights (123.9 hours). Stored at Palmdale 29 September 1966. Displayed at Museum of Flight, Seattle, WA, with D-21 (Article 510).
M-21	60-6941	135	First flight in May 1965. Successfully launched three drones Articles 503, 506, and 505). 95 flights (152.7 hours). Crashed 30 July 1966, following mid-air collision with D-21 (Article 504).

Table A4 D-21 (Total built: 38)

Model	Serial no.	Article no.	History
D-21 (D-21B)	N/A	501	Delivered as D-21 in December 1964. Used for fit checks and captive-carry flights on M-21. Modified to D-21B configuration and redelivered 27 July 1967. Accidentally dropped from B-52H during captive functional check flight on 28 September 1967. Crashed while still attached to booster. No mission flown. No camera.
D-21 (D-21B)	N/A	502	Delivered 2 November 1965. Modified to D-21B standard. Stored at Davis–Monthan AFB, Tucson, AZ. Later displayed at Beale AFB, Marysville, CA.
D-21	N/A	503	Delivered 2 November 1965. Launched from M-21 on 5 March 1966. Flew 150 nautical miles. Route not completed. Flight time 4.5 minutes. Inlet buzz at pushover affected pressure readings, causing false Mach-number value. Self-destructed. No camera. Hatch not recovered.
D-21	N/A	504	Delivered 26 April 1966. Launched from M-21 on 30 July 1966. Collided with M-21 immediately after launch, resulting in loss of both vehicles. No camera.
D-21	N/A	505	Delivered 6 April 1966. Launched from M-21 on 16 June 1966. Flew 1550 nautical miles. Route completed. Flight time 45.0 minutes. Cold battery caused loss of control at receiver switchover for descents. Self-destructed. No camera. Hatch not recovered.
D-21	N/A	506	Delivered 9 March 1966. Launched from M-21 on 27 April 1966. Flew 1120 nautical miles. Flight time 35.0 minutes. Route not completed because of hydraulic pump failure. Self-destructed. No camera. Hatch not recovered.
D-21 (D-21B)	N/A	507	Modified to D-21B standard. Delivered 13 October 1967. Launched from B-52 on 6 November 1967. Flew 134 nautical miles. Flight time 3.2 minutes. Route not completed because plastic inlet spike delaminated, causing inlet unstart and pitchover. Self-destructed. No camera. Hatch not recovered.
D-21 (D-21B)	N/A	508	Modified to D-21B standard. Delivered 2 November 1967. Launched from B-52 on 19 January 1968. Flew 280 nautical miles. Flight time 16.3 minutes. Route not completed. Lost right elevon control. Inlet unstarts during cruise with improved plastic spike. No camera. Hatch not recovered. Self-destructed.
D-21 (D-21B)	N/A	509	Modified to D-21B standard. Delivered 2 November 1967. Launched from B-52 on 2 December 1967. Flew 1430 nautical miles. Route not completed because of hydraulic failure. Equipped with nickel alloy spike. Self-destructed. No camera. Hatch not recovered.

(Continued)

Table A4   D-21 (Total built: 38) (*continued*)

Model	Serial no.	Article no.	History
D-21 (D-21B)	N/A	510	Modified to D-21B standard. No mission flown. Stored at Norton AFB, CA, in 1971. Transferred to Davis–Monthan AFB, AZ, 1976. Displayed at Museum of Flight, Seattle, WA, mated to M-21.
D-21 (D-21B)	N/A	511	Modified to D-21B standard. Delivered 13 April 1968. Launched from B-52 on 30 April 1968. Flew 150 nautical miles. Flight time 6.6 minutes. Route not completed. Tilted inlet unstarted, possibly because of plastic spike failure. Self-destructed. No camera. Hatch not recovered.
D-21 (D-21B)	N/A	512	Modified to D-21B standard. Delivered 29 April 1968. Launched from B-52 on 16 June 1968. Flew 2850 nautical miles. Flight time 86.0 minutes. Route completed. Equipped with metal spike. Achieved speed of Mach 3.25. Suffered unstarts during turns. Self-destructed. No camera. Hatch not recovered.
D-21 (D-21B)	N/A	513	Modified to D-21B standard. No mission flown. Stored at Norton AFB, CA, in 1971. Transferred to Davis–Monthan AFB, AZ, 1976.
D-21 (D-21B)	N/A	514	Modified to D-21B standard. Delivered 7 June 1968. Launched from B-52 on 1 July 1968. Flew 80 nautical miles. Flight time 3.2 minutes. Route not completed because of inertial coupling prior to booster separation. Self-destructed. No camera. Hatch recovered in water.
D-21 (D-21B)	N/A	515	Modified to D-21B standard. Delivered 1 December 1968. Launched from B-52 on 15 December 1968. Flew 2953 nautical miles. Flight time 92.86 minutes. Route completed. Self-destructed. Hatch and camera recovered, fair photos.
D-21 (D-21B)	N/A	516	Modified to D-21B standard. Delivered 31 July 1968. Launched from B-52 on 28 August 1968. Flew 78 nautical miles. Flight time 3.9 minutes. Route not completed because engine failed to accelerate to maximum power. Self-destructed. Camera and hatch not recovered.
D-21 (D-21B)	N/A	517	Modified to D-21B standard. Delivered 30 June 1969. Launched from B-52 on 9 November 1969. First operational mission. Flew more than 300 nautical miles. Flight time unknown. Drone disappeared following INS malfunction. Presumed self-destructed. Camera and hatch not recovered.
D-21 (D-21B)	N/A	518	Modified to D-21B standard. Delivered 18 November 1968. Launched from B-52 on 11 February 1969. Flew 161 nautical miles. Flight time 5.72 minutes. Route not completed because of flight control system failure. Self-destructed. Hatch and camera not recovered.

(*Continued*)

Table A4 D-21 (Total built: 38) (*continued*)

Model	Serial no.	Article no.	History
D-21 (D-21B)	N/A	519	Modified to D-21B standard. Delivered 27 January 1969. Launched from B-52 on 10 May 1969. Flew 2972 nautical miles. Flight time 92.5 minutes. Route completed. Self-destructed. Hatch and camera recovered.
D-21 (D-21B)	N/A	520	Modified to D-21B standard. Delivered 28 April 1969. Launched from B-52 on 10 July 1969. Flew 2937 nautical miles. Flight time 91.4 minutes. Self-destructed. Hatch and camera recovered, good photos.
D-21B	N/A	521	Delivered 13 August 1969. Launched from B-52 on 20 February 1970. Flew 2969 nautical miles. Flight time 92.6 minutes. Route completed. Unstarts during cruise. High payload vibration and temperatures. Self-destructed. Hatch and camera recovered, good photos.
D-21B	N/A	522	No mission flown. Stored at Norton AFB, CA, in 1971. Transferred to Davis–Monthan AFB, AZ, 1976. Delivered to NASA Dryden Flight Research Center on 1 June 1994. Borrowed by Northrop-Grumman. Returned to Dryden Flight Research Center in July 1998. Departed to Vandenberg AFB on 4 April 2007. Later transferred to Davis–Monthan AFB.
D-21B	N/A	523	Delivered 13 November 1969. Launched from B-52 on 16 December 1970. Flew 2448 nautical miles. Flight time 76.6 minutes. Route completed. Second operational mission. Self-destructed. Hatch and camera not recovered.
D-21B	N/A	524	No mission flown. Stored at Norton AFB, CA, in 1971. Transferred to Davis–Monthan AFB, AZ, 1976. Displayed at National Museum of the U.S. Air Force.
D-21B	N/A	525	No mission flown. Stored at Norton AFB, CA, in 1971. Transferred to Davis–Monthan AFB, AZ, 1976. Delivered to NASA Dryden Flight Research Center on 1 June 1994. Departed to Blackbird Airpark, Palmdale, CA, on 14 October 1994.
D-21B	N/A	526	Delivered 19 March 1970. Launched from B-52 on 4 March 1971. Flew 2935 nautical miles. Flight time 89.3 minutes. Route completed. Third operational mission. Drone crashed in the ocean and sank. Hatch and camera not recovered.
D-21B	N/A	527	Delivered 31 March 1970. Launched from B-52 on 20 March 1971. Flew 2935 nautical miles. Fourth and final operational mission. Flight time and distance unknown. Drone crashed or shot down. Hatch and camera not recovered.

(*Continued*)

Table A4 D-21 (Total built: 38) (*continued*)

Model	Serial no.	Article no.	History
D-21B	N/A	528	No mission flown. Stored at Norton AFB, CA, in 1971. Transferred to Davis–Monthan AFB, AZ, 1976. Displayed at Grissom Air Museum, Peru, IN.
D-21B	N/A	529	No mission flown. Stored at Norton AFB, CA, in 1971. Transferred to Davis–Monthan AFB, AZ, 1976. Delivered to NASA Dryden Flight Research Center on 2 June 1994. Borrowed by Northrop-Grumman. Returned to Dryden Flight Research Center in July 1998. Departed to Vandenberg AFB on 4 April 2007. Later transferred to Davis–Monthan AFB.
D-21B	N/A	530	No mission flown. Stored at Norton AFB, CA, in 1971. Transferred to Davis–Monthan AFB, AZ, 1976.
D-21B	N/A	531	No mission flown. Stored at Norton AFB, CA, in 1971. Transferred to Davis–Monthan AFB, AZ, 1976.
D-21B	N/A	532	No mission flown. Stored at Norton AFB, CA, in 1971. Transferred to Davis–Monthan AFB, AZ, 1976.
D-21B	N/A	533	No mission flown. Stored at Norton AFB, CA, in 1971. Transferred to Davis–Monthan AFB, AZ, 1976. Displayed at Pima Air Museum, Tucson, Ariz.
D-21B	N/A	534	No mission flown. Stored at Norton AFB, CA, in 1971. Transferred to Davis–Monthan AFB, AZ, 1976.
D-21B	N/A	535	No mission flown. Stored at Norton AFB, CA, in 1971. Transferred to Davis–Monthan AFB, AZ, 1976. Deaccessioned and returned to Defense Reutilization Management Office.
D-21B	N/A	536	No mission flown. Stored at Norton AFB, CA, in 1971. Transferred to Davis–Monthan AFB, AZ, 1976.
D-21B	N/A	537	No mission flown. Stored at Norton AFB, CA, in 1971. Transferred to Davis–Monthan AFB, AZ, 1976. Delivered to NASA Dryden Flight Research Center on 2 June 1994. Borrowed by Northrop–Grumman. Returned to Dryden Flight Research Center in July 1998. Stored at Edwards AFB North Base, CA, 10 May 2007 to July 2007. Displayed at March Field Museum, Riverside, CA.
D-21B	N/A	538	No mission flown. Stored at Norton AFB, CA, in 1971. Transferred to Davis–Monthan AFB, AZ, 1976. Displayed at Museum of Aviation, Robins AFB, CA.

Table A5 SR-71 (Total built: 32)

Model	Serial no.	Article no.	History
SR-71A	61-7950	2001	First flight 22 December, 1964. Used for Category I performance, stability and control, structural loads, and environmental testing. Damaged beyond repair 10 January 1967. Scrapped.
SR-71A	61-7951	2002	First flight 5 March 1965. Used for Category I system, subsystem and sensor development testing. Participated in joint USAF/NASA research program 1971 to 1978 under bogus designation "YF-12C." Stored at Palmdale in 1978. Displayed at Pima Air Museum, Tucson, AZ.
SR-71A	61-7952	2003	First flight 24 March 1965. Used for Category I systems and sensors testing. Crashed on 25 January 1966.
SR-71A	61-7953	2004	First flight 4 June 1965. Used for Category II performance, stability and control testing. Attained unofficial record cruising altitudes of 86,700 feet and 89,650 feet in 1968. Crashed 18 December 1969.
SR-71A	61-7954	2005	First flight 10 February 1966. Used for Category II systems testing. Damaged beyond repair 11 April 1969. Scrapped.
SR-71A	61-7955	2006	First flight 17 August 1965. Bailed to Lockheed as systems test aircraft. Used for maximum range tests during Category II testing. Served exclusively as test article. Last flight 24 January 1995. Displayed at Air Force Flight Test Center Museum, Edwards, CA.
SR-71B	61-7956	2007	First flight 18 November 1965. Assigned to 9SRW. Served as Air Force trainer until first retirement in 1990. Assigned to NASA in 1991. Last flight 19 October 1997. Stored at NASA Dryden Flight Research Center. Transferred to Kalamazoo Air Museum, Kalamazoo, MI, 2003.
SR-71B	61-7957	2008	First flight 10 December 1965. Trainer, used for handling-qualities testing during Category II. Assigned to 9SRW. Crashed on 11 January 1968.
SR-71A	61-7958	2009	First flight 15 December 1965. Used to set world absolute speed record of 2193 mph and 1000 km closed-course record of 2092 mph in July 1976. 2288.9 flight hours Last flight 23 February 1990 to Robins AFB, GA, for display in Museum of Aviation.
SR-71A	61-7959	2010	First flight 19 January 1966. Modified to BIG TAIL configuration in 1975 to increase sensor capacity. 866.1 flight hours Last Flight 29 October 1976. Displayed at Air Force Armament Museum, Eglin AFB, FL.

(Continued)

Table A5 SR-71 (Total built: 32) (*continued*)

Model	Serial no.	Article no.	History
SR-71A	61-7960	2011	First flight 9 February 1966. Flew 342 operational missions, more than any other SR-71. 1669.6 flight hours Last flight 27 February 1990. Displayed at Castle AFB, CA.
SR-71A	61-7961	2012	First flight 13 April 1966. Last flight 2 February 1977. 1601.0 flight hours Displayed at Cosmosphere and Space Center, Hutchison, KS.
SR-71A	61-7962	2013	First flight 29 April 1966. Set official altitude record of 85,068 feet on 28 July 1976. Last SR-71 to depart Kadena 21 January 1990. Placed in storage at Palmdale 14 February 1990. Selected for reactivation. Delivered to Imperial War Museum in Duxford, England, in 2001.
SR-71A	61-7963	2014	First flight 9 June 1966. 1604.4 flight hours Last flight 28 October 1976. Displayed at Beale AFB.
SR-71A	61-7964	2015	First flight 11 May 1966. 3373.1 flight hours Last flight 20 March 1990. Displayed at Strategic Air Command Museum, Ashland, NE.
SR-71A	61-7965	2016	First flight 10 June 1966. Crashed on 25 October 1967.
SR-71A	61-7966	2017	First flight 1 July 1966. 64.4 flight hours Crashed 13 April 1967.
SR-71A	61-7967	2018	First flight 3 August 1966. Used for maximum range tests during Category II. Assigned to 9SRW. Stored at Palmdale 14 February 1990. Reactivated in 1995. Assigned to DET 2, 9RW. Arrived at NASA Dryden Flight Research Center 14 October 1999 for storage. 2636.8 flight hours Returned to Air Force 28 November 2001 for display at 8th Air Force Museum at Barksdale AFB near Shreveport, LA.
SR-71A	61-7968	2019	First flight 10 October 1966. Set official endurance record of 15,000 miles in 10.5 hours on 26 April 1971. Last flight 12 February 1990. 2279.0 flight hours Delivered to DET 2, 9RW as dedicated reserve aircraft in 1998. Departed in October 1999. Displayed Virginia Aviation Museum, Richmond, VA.
SR-71A	61-7969	2020	First flight 18 October 1966. Crashed on 10 May 1970.
SR-71A	61-7970	2021	First flight 21 October 1966. Assigned to the 9SRW. Deployed to Kadena Air Base, Okinawa, Japan, from September 1968 to April 1969. 545.3 flight hours Crashed on 17 June 1970.
SR-71A	61-7971	2022	First flight 17 November 1966. Delivered to NASA Dryden 19 March 1990. Departed to Lockheed Palmdale 12 January 1995. Assigned to DET 2, 9RW in April 1995. Last flight 1998. 3512.4 flight hours Returned to Dryden 14 October 1999. Departed to Evergreen Aviation Museum in McMinnville, OR.

(*Continued*)

Table A5 SR-71 (Total built: 32) *(continued)*

Model	Serial no.	Article no.	History
SR-71A	61-7972	2023	First flight 12 December 1966. Inadvertently flown on first international sortie 2 July 1967 when inertial navigation system failure caused incursion into Mexican airspace. Set New York to London speed record 1 September 1974 (3461 miles in 65 minutes). First SR-71 deployed to DET 4, Mildenhall, England, in April 1976. Last flight 6 March 1990 (set four world speed records, including west coast to east coast 2404 miles in just under 68 minutes). 2801.1 flight hours. Aircraft displayed at Udvar-Hazy Gallery, National Air and Space Museum Washington, D.C.
SR-71A	61-7973	2024	First flight 8 February 1967. 1729.9 flight hours Airframe overstressed during air show demonstration at Royal Air Force, Mildenhall, England, in May 1987. Last flight 21 July 1987. Displayed at Blackbird Airpark, Palmdale, CA.
SR-71A	61-7974	2025	First flight 16 February 1967. Crashed on 21 April 1989.
SR-71A	61-7975	2026	First flight 13 April 1967. 2854.0 flight hours Last flight 28 February 1990. Displayed at March Field Museum, Riverside, CA.
SR-71A	61-7976	2027	First flight 13 May 1967. First operational SR-71 mission from Kadena 21 March 1968. Last flight 27 March 1990. Displayed at National Museum of the U.S. Air Force, Dayton, OH.
SR-71A	61-7977	2028	First flight 6 June 1967. Crashed during takeoff on 10 October 1968.
SR-71A	61-7978	2029	First flight 5 July 1967. First SR-71 deployed to Kadena 8 March 1968. Crashed during landing on 20 July 1972.
SR-71A	61-7979	2030	First flight 10 August 1967. Flew the first three of nine sorties from the U.S. to the 1973 Middle East during the Yom Kippur War. 3321.7 flight hrs. Last flight 6 March 1990. Displayed at Lackland AFB, Texas.
SR-71A	61-7980	2031	First flight 25 September 1967. Last production SR-71. Transferred to NASA Dryden in February 1990. Last flight 9 October 1999. Displayed at NASA Dryden Flight Research Center, Edwards, CA.
SR-71C	61-7981	2000	Trainer built from forward fuselage of engineering test article and aft fuselage of YF-12A (1001). First flight 14 March 1969. 556.4 flight hrs. Last flight 11 April 1976. Displayed at Hill AFB, Utah.

Appendix B

BLACKBIRD TIMELINE

This section provides a brief chronology of significant events and milestones in the history of the Blackbirds.

1955
August 4 — First flight of Lockheed U-2 (Project AQUATONE)

1956
May 1 — USAF issues a contract to Lockheed for development of the CL-400 hydrogen-powered aircraft (Project SUNTAN).

1957
March — Kelly Johnson recommends termination of SUNTAN.
August — Project GUSTO initiated to replace the U-2 with a high-performance aircraft capable of evading radar detection.
December 24 — First Pratt and Whitney JTD11D-20 (J58) engine run.

1958
April 21 — Kelly Johnson initiates Archangel design development. First studies of a Mach 3.0 cruise airplane having a 4,000 nautical mile range at altitudes over 90,000 feet.
April 23 — Johnson completes pencil sketch of Archangel 1 (A-1) configuration.
July 23 — Johnson presents GUSTO 2A design to Richard Bissell at CIA.
August — Johnson develops A-2 design.
November — Johnson completes A-3 design drawing.
December — Johnson develops A-4, A-5, and A-6 designs.

1959
January — Sacrificing radar-cross-section reduction for increased performance, Johnson designed the A-7, A-8, and A-9 configurations.
February — SUNTAN terminated. Johnson submits the A-10 design.
March — Johnson refines his configuration as the A-11.
March 31 — Construction of full-scale radar test mock-up begins.
July 3 — Lockheed receives a program extension to improve radar-cross-section reduction.
August 20 — Lockheed submits A-12 configuration.
August 28 — CIA accepts A-12 design proposal under Project OXCART, subject to proof of low radar cross section by 1 January 1960. Project GUSTO is terminated.

November 9	Full-scale radar-cross-section model completed.
November	Early A-12 configuration model undergoes radar-cross-section tests.

1960

January 1	Johnson provided CIA with data proving the radar cross section of the A-12 design met all requirements.
January 26	CIA gives Lockheed initial go-ahead for production of static test article and 10 or 12 airplanes.
January 30	CIA approves full funding for A-12 production.
February	OXCART pilot selection begins.
March 16	Johnson proposes USAF interceptor variant (AF-12).
April 5	First titanium parts are completed.
October	USAF authorizes production of AF-12 prototype under Project KEDLOCK.

1961

January	Johnson proposes R-12 reconnaissance variant for USAF.
May 31	AF-12 mock-up review takes place at Lockheed's Burbank, CA, plant.
September	Because the J58 engine would not be available for initial A-12 flights, Johnson proposed temporarily installing less powerful J75 engines.

1962

February 26	First A-12 (Article 121) shipped to test site, arriving two days later.
February	Johnson proposes AQ-12 drone for launch from OXCART-type aircraft.
April 25	A-12 completes high-speed taxi tests and briefly becomes airborne.
April 26	A-12 completes first planned flight, powered by J75 engines.
April 30	Official first flight of A-12, in front of distinguished guests.
May 4	Article 121 completes first supersonic flight (Mach 1.10)
June 4	USAF reviews R-12 mock-up.
June 26	Article 122 arrives at the test site for pole-mounted radar-cross-section tests.
July 30	Pratt and Whitney J58 engine completes ground testing.
October 5	Article 121 flies with J58 engine in right nacelle and J75 in left nacelle.
October 10	CIA authorizes Lockheed to develop AQ-12 drone under Project TAGBOARD.
November	A-12T trainer (Article 124) arrives at the test site.
December 17	Article 125 arrives at the test site. Last airframe to be initially equipped with J75 engines.
December 28	Lockheed signs contract to build six R-12 airframes.

1963

January 15	First flight of Article 122. First flight of Article 121 equipped with two J58 engines.
January 22	First flight of A-12T.
March 9	First flight of Article 121 with two J58 engines.
March	Article 126 becomes first A-12 to complete its maiden flight equipped with J58 engines. Article 127 makes first flight.
May 24	Article 123 crashes during test flight. First loss of a Blackbird.
May 31	AF-12 mock-up review held in Burbank.
June	AQ-12 drone (Article 501) is mated to M-21 (Article 134) for fit check. First flight of Article 128.
July 20	First A-12 flight to exceed Mach 3.0 (Article 121).
August 7	First flight of AF-12 (Article 1001).

APPENDIX B

October 1	AQ-12 design finalized and redesignated D-21.
November	A-12 achieves design cruise speed (Mach 3.2) and altitude of 78,000 feet.

1964

February 3	A-12 cruises at Mach 3.2 and 83,000 feet for 10 minutes.
February 29	President Lyndon B. Johnson announces existence of AF-12 as "A-11." Aircraft is soon redesignated YF-12A.
April 16	First missile separation tests from YF-12A.
May 6	Johnson briefed General Dynamics representatives on Lockheed's experience with engines, exhaust ejectors, and high-speed cruise.
July 24	President Johnson announces R-12 as "SR-71."
October 23	A-12 (Article 128) flew 4,500 nautical miles with two refuelings. The mission included multiple legs at Mach 3.0 and 80,000 feet.
October 29	First SR-71 (Article 2001) delivered to Lockheed's Palmdale, CA, facility for final assembly.
December 18	SR-71 begins engine runs.
December 21	First SR-71 taxi tests.
December 22	First flights of SR-71 (Article 2001) and MD-21 mated combination (Article 134/Article 501).

1965

January 9	YF-12A flies at Mach 3.2 for 5 minutes.
January 27	A-12 (Article 129) participates in SILVER JAVELIN cruise endurance test at or above mach 3.1 for 1.25 hours.
March 18	First firing of Hughes GAR-9 missile from YF-12A.
May 1	YF-12A sets several official world speed and altitude records including maximum speed of 2,070 mph (Mach 3.14) and maximum altitude of 80,257 ft.
May 8	A-12 sets unofficial record speed of 2,171 mph (Mach 3.32).
June 1	SR-71/F-12 Test Force is established at Edwards AFB.
August 14	A-12 sets unofficial altitude record of 90,000 feet.
September 28	GAR-9 missile fired from YF-12A at Mach 3.2 and 75,000 ft altitude at a target 36 miles away. The missile passed within 6 feet of the drone.
November 5	Under Project SKYLARK, several A-12 aircraft are prepared for emergency operational capability (Cuban overflights).
November 20	A-12 completes 6.33-hour mission, the longest to date.
November 30	Three A-12 airplanes declared ready for BLACK SHIELD reconnaissance missions.
December 18	First flight of SR-71B trainer.

1966

January 7	First SR-71B trainer delivered to USAF.
January 25	First loss of SR-71 (Article 2003), during a test flight.
February 11	USAF Staff Crew #1 completes initial SR-71 qualification.
March 5	First D-21 free flight. Article 501 launched from M-21 (Article 135).
April 29	Second batch of 15 D-21 drones ordered. Kelly Johnson proposes modification to launch drones from a B-52.
May 24	First SR-71A delivered to Beale AFB, CA.
July 30	Accident during fourth D-21 launch results in loss of M-21 (Article 135) and drone (Article 504), and fatal injuries to launch control officer.
August 14	First loss of YF-12A, due to landing accident.

September	Fischer–Bennington Report from the Bureau of the Budget recommends OXCART airplanes be stored as surplus and the reconnaissance mission be transferred entirely to Strategic Air Command SR-71 aircraft.
December 12	First of two B-52H aircraft arrived at Lockheed for modification to drone launch platforms under Project SENIOR BOWL.
December 21	A-12 completes 10,200-mile sortie in 6 hours.
December 23	Decision is made to terminate Project OXCART by 01 June 1968.

1967

January 10	Schedule for storage of A-12 airframes at AF Plant 42 is completed.
January	In a desperate effort to save the OXCART program, Johnson proposes converting half the SR-71 fleet to bombers to eliminate perceived surplus of reconnaissance aircraft.
April 17	SR-71 completes 14,000-mile flight. Crew receives FAI Gold Medal.
May	Secretary of Defense directs that SR-71 assume duties previously assigned to OXCART, but plans continue for Operation BLACK SHIELD overflights of southeast Asia.
May 22	First of three A-12 aircraft is deployed to Kadena, Japan, for Operation BLACK SHIELD. Two others follow.
May 29	BLACK SHIELD unit declared operational.
May 31	First operational A-12 mission accomplished over North Vietnam.
July 2	First international SR-71 sortie inadvertently flown when aircraft on training mission flew into Mexican airspace due to INS failure.
September 28	D-21B (Article 501) is accidentally dropped from B-52H during captive test flight.
October 30	A-12 struck by small piece of shrapnel from surface-to-air missile over North Vietnam resulting in only combat damage ever sustained by a Blackbird.
November 3	A-12 and SR-71 tested against each other in Project NICE GIRL fly-off over the Mississippi Valley. Results were inconclusive.
November 6	First planned launch of D-21B from B-52H.
December	KEDLOCK (F-12) program terminated.

1968

January 5	Lockheed receives notice that USAF is canceling its order for the F-12B and terminating YF-12A test operations.
January 26	First A-12 overflight of North Korea.
February 1	Last YF-12A test flight.
February 5	USAF instructs Lockheed to destroy tooling for A-12/YF-12/SR-71.
February 14	Successful mission to locate USS Pueblo, seized by North Korea.
March 8	First SR-71 deployed to Kadena.
March 21	First operational SR-71 mission over North Vietnam.
May 8	Last A-12 mission over North Korea.
June 4	Fifth and final A-12 loss when Article 129 crashed in South China Sea.
June 7	Remaining BLACK SHIELD aircraft redeploy to United States.
June 21	Last A-12 placed in Storage at AF Plant 42.
June 26	BLACK SHIELD pilots awarded CIA Intelligence Star for Valor.
November 2	9th Strategic Reconnaissance Wing at Beale AFB receives Air Force Outstanding Unit Award, the first of many.
December	SR-71 Category II operational testing is completed.

1969
June 5 NASA and USAF sign agreement to conduct a joint research and test program with two YF-12A aircraft.
November 9 First operational D-21B mission completed, but camera not recovered.
December 11 First flight of joint NASA-USAF YF-12 research program.

1971
March 20 Fourth and final operational D-21B flight; camera not recovered.
April 26 SR-71 completes 15,000-mile sortie in 10.5 hours. Crew later received the McKay Trophy and Harmon International Aviator Award.
June 24 Second and final YF-12A loss; crashed during test flight.
July 17 TAGBOARD/SENIOR BOWL program cancelled.

1973
October 12 SR-71 accomplished first CONUS-based sortie over the Middle East under Operation GIANT REACH.

1974
September 1 First SR-71 deployment to United Kingdom set a world record: New York to London (3,490 nautical miles) in less than 2 hours.
September 13 SR-71 set world record London to Los Angeles: 5,645 miles in 3 hours 47 minutes.

1975
January Clarence "Kelly" Johnson retires from Lockheed, but remains as a consultant.
January 17 Ben Rich becomes chief of the Skunk Works.
April SR-71 aircraft deployed to Mildenhall, England.
December 3 First flight of SR-71 Big Tail configuration.

1976
April First operational missions flown from Mildenhall.
July 27 SR-71 sets world closed-course speed record (2,092 mph).
July 28 SR-71 sets world straight course speed record of 2,193 mph (Mach 3.32) and world altitude record of 85,069 feet.
July 4 First SR-71 BigTail flight at Mach 3.0.
October 29 Last Big Tail flight.

1979
October 31 Last flight of joint NASA-USAF YF-12 research program.
November 7 Last surviving YF-12A makes final flight.

1980
August Honeywell starts conversion of SR-71 analog flight and inlet control system to digital automatic flight and inlet control system.

1982
January 15 SR-71B (Article 2007) completes its 1000th sortie.

1984
November 7 First SR-71 overflights of Central America.

1985
January 24 Last flight (722nd sortie) of SR-71 dedicated test airframe (Article 2006).

January 28	Article 2006 used for runway roughness evaluation taxi tests.
November	Honeywell completes conversion from analog to digital controls.

1989

April 21	Crash near Philippines is 12th and final SR-71 loss.
October 1	USAF SR-71 operations are suspended except for minimum necessary to retain proficiency.
November 22	SR-71 program is officially terminated.

1990

January 21	Last SR-71 redeployed from Kadena to United States.
January 26	SR-71 decommissioning ceremony held at Beale AFB.
February	SR-71A (Article2031) delivered to NASA for flight research.
March 6	SR-71 sets four world speed records between Los Angeles and Washington, D.C., including flying 2,404 miles in less than 68 minutes.
December 22	Clarence "Kelly " Johnson dies at age 80.
December 23	Ben Rich retires from Lockheed.

1991

July 25	SR-71B delivered to NASA for flight research.
October	Marta Bohn-Meyer qualifies as first female SR-71 crew member.

1993

March 9	First flight for NASA SR-71 research program.

1994

September 28	Congress votes to allocate $100 million for reactivation of three SR-71 aircraft for USAF.

1995

April 26	First flight of SR-71A for USAF reactivation program.
June 28	First reactivated SR-71 officially returns to USAF inventory.
August 29	First new USAF crew declared mission ready.

1996

January 30	First newly operational SR-71 deployed to Edwards AFB.
February 1	Second newly operational SR-71 deployed to Edwards AFB.
April 15	Deputy secretary of defense John White directs USAF to SR-71 because of conflicting language in Section 304 of the National Security Act of 1947, and Section 102 of the Intelligence Authorization Act for FY-96.
April 16	SR-71 flight operations are suspended.
September 21	House and Senate Appropriation committees have agreed to fund the Air Force's 2 operational SR-71s for Fiscal Year 1997.

1997

January 1	SR-71 aircraft and crews at Edwards are declared operational.
October 10	Last USAF SR-71 flight.
October 15	President kills SR-71 funding with line-item veto.
October 19	Last flight of SR-71B by NASA.

1999

June 30	USAF SR-71 program is ordered shut down.
October 9	Final flight of SR-71A by NASA. Last Blackbird flight ever.

NOTES

[1] Crickmore, Paul F., *Lockheed Blackbird—Beyond the Secret Missions*, Osprey Publishing, Oxford, England, UK, 2004.

[2] Goodall, James, and Jay Miller *Lockheed's SR-71 'Blackbird' Family*, Midland Publishing, Hinckley, England, UK, 2002.

[3] Whittenbury, John R., "From Archangel to Oxcart: Design Evolution of the Lockheed A-12, First of the Blackbirds," unpublished briefing slides, Lockheed Martin Aeronautics Co., Palmdale, CA, 2003.

[4] "The U-2's Intended Successor: Project Oxcart, 1956–1968," Central Intelligence Agency, Langley, VA, declassified 1994.

[5] Johnson, Clarence L., *History of the Oxcart Program* (SP-1362), Lockheed Aircraft Corp., Advanced Development Projects, Burbank, CA, 1968.

[6] Goodall and Miller, *Lockheed's SR-71 'Blackbird' Family*.

[7] Johnson, Clarence L., "Proposal for a High Speed Reconnaissance Aircraft," Report #1, Lockheed Corp., Burbank, CA, Dec. 1958.

[8] Whittenbury, "From Archangel to Oxcart."

[9] Ibid.

[10] Johnson, Clarence L., *History of the Oxcart Program*, and *Proposal for a Lightweight Reconnaissance Aircraft* (SP-108), Lockheed Aircraft Corp., Advanced Development Projects, Burbank, CA, 1958.

[11] Johnson, Clarence L., "Archangel Log," Lockheed Corp., Burbank, CA, released in abridged form, April 1993.

[12] Whittenbury, "From Archangel to Oxcart."

[13] Johnson, "Archangel Log."

[14] Whittenbury, "From Archangel to Oxcart."

[15] *A-12 Technical Manual MA12-2-1—Airframe*, Lockheed Aircraft Corp., Burbank, CA, Feb. 1966.

[16] *Lockheed SR-71 Supersonic/Hypersonic Research Facility Researcher's Handbook*, Vol. II, Technical Description, Lockheed Advanced Development Co., Palmdale, CA, 1995.

[17] *A-12 Technical Manual MA12-2-1—Airframe*.

[18] Ibid.

[19] Crickmore, *Lockheed Blackbird—Beyond the Secret Missions*.

[20] *A-12 Technical Manual MA12-2-1—Airframe*.

[21] *Lockheed SR-71 Supersonic/Hypersonic Research Facility Researcher's Handbook*, Vol. II, Technical Description.

[22] Crickmore, *Lockheed Blackbird—Beyond the Secret Missions*.

[23] Johnson, *History of the Oxcart Program*.

[24] Crickmore, *Lockheed Blackbird—Beyond the Secret Missions*.

[25] Johnson, "Archangel Log."

[26] Goodall and Miller, *Lockheed's SR-71 'Blackbird' Family*.

[27] Johnson, "Archangel Log."
[28] Crickmore, *Lockheed Blackbird—Beyond the Secret Missions*.
[29] Merlin, Peter W., *Mach 3+: NASA/USAF YF-12 Flight Research, 1969–1979*, NASA SP-2001-4525, NASA Headquarters, Washington, DC, 2002.
[30] Crickmore, *Lockheed Blackbird—Beyond the Secret Missions*.
[31] Merlin, *Mach 3+: NASA/USAF YF-12 Flight Research, 1969–1979*.
[32] Crickmore, *Lockheed Blackbird—Beyond the Secret Missions*.
[33] Angerman, G. J., *Structural Criteria and Design Loads, M-21* (SP-727), Lockheed Aircraft Corp., Advanced Development Projects, Burbank, CA, 1966.
[34] Crickmore, *Lockheed Blackbird—Beyond the Secret Missions*.
[35] *D-21 Technical Manual* (SP-790), Lockheed Aircraft Corp., Advanced Development Projects, Burbank, CA, 1971.
[36] Goodall and Miller, *Lockheed's SR-71 'Blackbird' Family*.
[37] *Technical Manual, Field Maintenance Instructions, Ramjet Engine Model No. MA20S-4* (MA20-R-3), The Marquardt Corp., Van Nuys, CA, 1965.
[38] Goodall and Miller, *Lockheed's SR-71 'Blackbird' Family*.
[39] Ibid.
[40] Crickmore, *Lockheed Blackbird—Beyond the Secret Missions*.
[41] Goodall and Miller, *Lockheed's SR-71 'Blackbird' Family*.
[42] Ibid.
[43] Johnson, Clarence L., "Development of the Lockheed SR-71 Blackbird," *Lockheed Horizons*, No. 9, Winter 1981/82, Lockheed Corp., Burbank, CA, 1981, pp. 2–18.
[44] Ibid.
[45] Johnson, "Archangel Log."
[46] Johnson, "Development of the Lockheed SR-71 Blackbird."
[47] Johnson, Clarence L., "Some Development Aspects of the YF-12A Interceptor Aircraft," *Journal of Aircraft*, Vol. 7, No. 4, AIAA, 1970, pp. 355–359.
[48] Johnson, "Development of the Lockheed SR-71 Blackbird."
[49] Johnson, "Some Development Aspects of the YF-12A Interceptor Aircraft."
[50] *Technical Manual SR-71-2-2, Organizational Maintenance—Airframe Group: SR-71, SR-71B, and SR-71C Aircraft*, U.S. Air Force, 1987.
[51] Johnson, "Development of the Lockheed SR-71 Blackbird."
[52] Merlin, Peter W., "SR-71 Blackbird," *Advanced Processes and Materials*, No. 161(5), ASM International, May 2003. Most of the material in the second half of this chapter was derived from the aforementioned article. Primary sources included: *A-12 Technical Manual MA12-2-2—Airframe*, 1966; *Technical Manual SR-71-2-2, Organizational Maintenance—Airframe Group: SR-71, SR-71B, and SR-71C Aircraft*, 1987; and *Lockheed's SR-71 'Blackbird' Family* by James Goodall and Jay Miller, Aerofax/Midland Publishing, 2002.
[53] Alitzer, John, "SR-71 Structures and Materials," part of "Case Studies in Engineering: The SR-71 Blackbird," Course Ae107, presented at the Graduate Aeronautical Lab., California Inst. of Technology, Pasadena, CA, April–May 1990.
[54] Merlin, "SR-71 Blackbird," *Advanced Processes and Materials*.
[55] Ibid.
[56] Johnson, "Some Development Aspects of the YF-12A Interceptor Aircraft."
[57] Johnson, "Development of the Lockheed SR-71 Blackbird."
[58] Crickmore, *Lockheed Blackbird—Beyond the Secret Missions*.
[59] *F-12 Emergency Ground Procedures Manual*, Change 1, June 1965.
[60] *SR-71A Flight Manual*, T.O. SR-71A-1, Sec. 6, "Flight Characteristics," Oct. 1986.
[61] Ibid.
[62] *A-12 Utility Flight Manual*, T.O. A12-1, Sec. 6, "Flight Characteristics," Sept. 1965.

[63] Meyer, J. E., J. R. McMaster, and R. L. Moody, H*andling Qualities of the SR-71*, SP-508, Lockheed Advanced Development Co., Burbank, CA, Oct. 1964, revised July 1967.

[64] Abrams, Richard, William A. Lusby, Jr., Mervin L. Evenson, and William L. Skliar, *SR-71A Category II Performance Tests*, 69-AFFTC-39459, Air Force Flight Test Center, Edwards, CA, March 1970.

[65] Bennett, Frenchy D., and Robert L. Stephens, (eds.), *SR-71 Final Report, Category II Flight Test Program—1 July 1965 to 30 June 1967*, 67-AFFTC-18546, Air Force Flight Test Center, Edwards, CA, July 1970.

[66] Allender, James R., Ernest L. Pyne, Mervin L. Evenson, and William L. Skliar, *SR-71A Category II Stability and Control Tests*, Air Force Flight Test Center, Edwards, CA, July 1970.

[67] Bennett et al.

[68] Crickmore, *Lockheed Blackbird—Beyond the Secret Missions*.

[69] Bennett et al.

[70] *Handling Qualities of the SR-71*.

[71] Abrams et al. and Bennett et al.

[72] Bennett et al.

[73] Abrams et al. and Crickmore, *Lockheed Blackbird—Beyond the Secret Missions*.

[74] *A-12 Utility Flight Manual*.

[75] Abrams et al.

[76] Crickmore, *Lockheed Blackbird—Beyond the Secret Missions*.

[77] Ibid.

[78] "OXCART A-12 Aircraft Experience Data and Systems Reliability," BYE-8725-68, Central Intelligence Agency, Washington, DC, Jan. 15, 1968.

[79] Allender et al.

[80] Urie, David, "Case Studies in Engineering: The SR-71 Blackbird," Course Ae107, presented at the Graduate Aeronautical Lab., California Inst. of Technology, Pasadena, CA, April–May 1990.

[81] Allender et al.

[82] Urie, David, "Case Studies in Engineering: The SR-71 Blackbird."

[83] Matranga, Gene, and William J. Fox, "YF-12A Development and Operational Experience," unpublished paper presented at the Supercruiser Conference, Wright–Patterson AFB, OH, Feb. 17–20, 1976, p. 3; part of NASA Dryden Flight Research Center Historical Reference Collection.

[84] Merlin, Peter W., *Mach 3+: NASA/USAF YF-12 Flight Research, 1969–1979*, NASA SP-2001-4525, NASA Headquarters, Washington, DC, 2002.

[85] Crickmore, *Lockheed Blackbird—Beyond the Secret Missions*.

[86] Ibid.

[87] *Lockheed SR-71 Supersonic/Hypersonic Research Facility Researcher's Handbook*, Vol. II.

[88] Crickmore, *Lockheed Blackbird—Beyond the Secret Missions*.

[89] *Lockheed SR-71 Supersonic/Hypersonic Research Facility Researcher's Handbook*, Vol. II.

[90] Memorandum from Gen. James T. Stewart regarding YF-12A speed trials, Central Intelligence Agency, 1964.

[91] Memorandum from Director of Central Intelligence John A. McCone regarding YF-12A speed trials discussion at Lockheed, Central Intelligence Agency, Aug. 18, 1964.

[92] Memorandum to Director of Central Intelligence John A. McCone from Col. Jack C. Ledford, assistant director for Special Activities (approved by Albert D. Wheelon, deputy director, Directorate of Science and Technology), Central Intelligence Agency, Aug. 20, 1964.

[93] "OXCART A-12 Aircraft Experience Data and Systems Reliability."

[94] Crickmore, *Lockheed Blackbird—Beyond the Secret Missions*.

[95] Abrams et al.

[96] Byrnes, Donn A., and Kenneth D. Hurley, *Blackbird Rising: Birth of an Aviation Legend*, Sage Mesa Publications, Los Lunas, NM, 1999.

[97] "OXCART Development Summary and Progress (1 Oct. 1966–31 Dec. 1966)," Tab A, Sec. 2, p. 3, Central Intelligence Agency, Langley, VA, n.d.

[98] Memorandum for Director of Special Activities from Assistant for Programs, Research and Development (Special Activities), regarding "Comments to W.R. Thomas III Memorandum to the Director, Bureau of Budget," Central Intelligence Agency, July 27, 1966.

[99] *Lockheed SR-71 Supersonic/Hypersonic Research Facility Researcher's Handbook*, Vol. II.

[100] Graham, *SR-71 Revealed: The Inside Story*, Motorbooks International, Osceola, WI, 1996.

[101] Memorandum for Director of Special Activities from Assistant for Programs, Research and Development (Special Activities), regarding "Comments to W.R. Thomas III, Memorandum to the Director, Bureau of Budget," Central Intelligence Agency, July 27, 1966.

[102] Graham, *SR-71 Revealed: The Inside Story*.

[103] "OXCART Development Summary and Progress (1 Oct. 1966–31 Dec. 1966)."

[104] Crickmore, *Lockheed Blackbird—Beyond the Secret Missions*.

[105] Memorandum for John L. McLucas, director, National Reconnaissance Office, from B/Gen. John E. Kulpa Jr., regarding "OXCART and SR-71 Considerations," Washington, DC, Sept. 1973.

[106] "Project Programs—Part II," briefing material from the center director's monthly Projects Review meeting, NASA Flight Research Center, Edwards, CA, March 1965, NASA Dryden Flight Research Center Historical Reference Collection.

[107] Memorandum from R. L. Miller to L. M. Bohanan regarding "Proposed Instrumentation for 2004 Follow On Tests," 17 June 1968, NASA Dryden Flight Research Center Historical Reference Collection. Note: 2004 is the Lockheed article number for SR-71A, serial 61-7953, the fourth SR-71 off the production line.

[108] Kock, Berwin, "Overview of the NASA YF-12 Program," presented during the NASA YF-12 Experiments Symposium held at Hugh L. Dryden Flight Research Center, Edwards, CA, Sept. 13, 1978, NASA Dryden Flight Research Center Historical Reference Collection.

[109] "Memorandum of Understanding, USAF-NASA Research and Development Program," 5 June 1969, NASA Dryden Flight Research Center Historical Reference Collection.

[110] Hallion, Richard P., *On The Frontier: Flight Research at Dryden, 1946–1981*, NASA SP-4303, Washington, DC, 1984, p. 191.

[111] Kock, "Overview of the NASA YF-12 Program."

[112] Miller, Jay, *Lockheed Skunk Works—The First Fifty Years*, Aerofax, Arlington, TX, 1993, p. 133.

[113] Gene Matranga interview by Peter Merlin, Lancaster, CA, 30 May 2000, transcript in the Dryden Flight Research Center Historical Reference Collection.

[114] Jenkins, Jerauld M., and Robert D. Quinn, "A Historical Perspective of the YF-12A Thermal Loads and Structures Program," NASA Dryden Flight Research Center, Edwards, CA, NASA TM-104317, May 1996, pp. 8–11.

[115] Ibid., p.12.

[116] Ibid., p.13.

[117] Albers, James A., *Status of the NASA YF-12 Propulsion Research Program*, NASA TM X-56039, March 1976. Reynolds number, named after Osborne Reynolds, is a nondimensional parameter equal to the product of the velocity of, in this case, an airplane passing through a fluid (air in this case), the density of the fluid and a representative length, divided by the fluid's

viscosity. In shorthand, this is the inertial force divided by the viscous forces for the mass of air acted upon by the vehicle. Among other uses, it served to compare data from wind-tunnel models with that from full-sized airplanes or components. The Reynolds number was not determined solely by the viscosity of the air. The YF-12, for example, would have a much larger Reynolds number when flying through air at a given altitude, location, and time than would a small model simply because of the difference in size and the amount of air displaced. Furthermore, the Reynolds number would be much larger at the rear of a flight vehicle than at the front.

[118] Ibid., p. 4.

[119] Two types of engines have been used in the SR-71 aircraft. The early type (referred to as the J-engine), also used in the YF-12A, incorporated fixed compressor inlet guide vanes and had a maximum afterburner thrust rating of 32,500 pounds at sea-level standard-day conditions. An improved configuration (the K-engine) incorporated two-position compressor inlet guide vanes. The vanes were automatically positioned axially below Mach 1.9 to provide increased airflow and increased thrust rating. Above Mach 1.9, the vanes moved to a cambered position, and the engine provided thrust equivalent to the J-engine. The K-engine had a maximum afterburner thrust rating of 34,000 pounds at sea-level standard-day conditions. Metal inlet spikes were originally used on some of the Blackbirds. A plastic spike was later incorporated into production aircraft to reduce the vehicle's radar cross section. By October 1974, the improved (K-type) engines were installed in the YF-12A. NASA conducted no propulsion research with the K-engines.

[120] "Project Activity Guide (Accomplishments)," dated June 2, 1971, and June 17, 1971, list all three aircraft indicating that the YF-12C was not simply a replacement for Article 1003, which crashed on 24 June 1971. Part of NASA Dryden Historical Reference Collection.

[121] "History of Flight," YF-12A Mishap Report, AF Form 711, Part 11, July 16, 1971.

[122] A phugoid occurs when an aircraft's airspeed or pitch attitude is disturbed from its trimmed equilibrium condition. During flight, the pilot trims the aircraft to a desired angle of attack (pitch attitude) that can then be disturbed by additional pilot input or natural air turbulence. The airplane's tendency to return to its trimmed attitude is so strong that it generally returns too quickly and overshoots. The oscillations tend to die out after a few cycles, and the aircraft returns to its trimmed condition.

[123] "Final Report for the CAPA/YF-12 Central Airborne Performance Analyzer, Phase II: The Study of In-Flight Real Time Diagnostics," Honeywell, Minneapolis, MN, Rept. W8340-FR, Nov. 8, 1974, pp. 1-1–4-2.

[124] Lift-to-drag ratio is the value of the total aerodynamic force acting on a body (such as an aircraft or wing) divided by the retarding force that acts on that body as it moves through a gaseous fluid.

[125] "YF-12 Runway Response," Jim McKay notebook. Part of NASA Dryden Historical Reference Collection. The original program, designed by NASA Langley for use with rigid bodies, was called Takeoff and Landing Analysis (TOLA). After McDonnell Douglas Astronautics Corporation programmers modified it with a flexible-body option, it was renamed Flexible Aircraft Takeoff and Landing Analysis (FATOLA). Jim McKay refers to it simply as TOLA throughout his notes, but it is more properly called FATOLA as used for the YF-12.

[126] Ibid.

[127] Ground effect is an increase in the lift of an aircraft operating close to the ground caused by reaction between high-velocity downwash from its wing and the ground.

[128] The air load-stroke curve is a function of landing-gear strut stroke and strut loads. The stroke (length of extension/compression of the gear strut) is a function of maximum touchdown load, impact velocity, stroking efficiency, and tire deflection.

[129] Gamon, Max A., "Testing and Analysis of Dual-Mode Adaptive Landing Gear, Taxi Mode Test System for YF-12A," Lockheed, NASA CR-144884, Washington, DC, Sept. 1979.

[130] Thompson, Milton O., "DFRC Orbiter Landing Investigation Team—Final Presentation," briefing at NASA Dryden Flight Research Center, Edwards, CA, 17 Aug. 1978. Part of NASA Dryden Flight Research Center Historical Reference Collection.

[131] Teper, Gary L., Richard J. DiMarco, and Irving L. Ashkenas, "Analysis of Shuttle Orbiter Approach and Landing Conditions," Systems Technology, Inc., Technical Report TR-1137-1, Hawthorne, CA, Jan. 1980, pp. 17–26.

[132] Combs, Henry G., and John R. McMaster, "Feasibility Study on Launching a NASA HT-4 Hypersonic Drone from a Lockheed YF-12C Airplane," Lockheed Rept. SP-1779, July 30, 1971. Part of NASA Dryden Flight Research Center Historical Reference Collection.

[133] Miller, *Lockheed's Skunk Works—The First Fifty Years*, pp. 137, 138.

[134] *Basic Research Review for the NASA OAST Research Council*, RTOP-761-74-01, Robert D. Quinn, Aug. 1971, p. 15; RTOP-501-06-05, RTOP-501-06-08, RTOP-770-18-10, RTOP 501-06-05, Edwin J. Salzmann, Aug. 7, 1972, and Aug. 20, 1974, pp. 9, 10; RTOP-505-06-31, Sheryll Goecke Powers, Oct. 20, 1975, p. 9.

[135] "Report of Investigation," Memorandum from YF-12 Ventral Incident Investigation Board to Director of NASA Hugh L. Dryden Flight Research Center, p. 5, April 25, 1975. Part of NASA Dryden Flight Research Center Historical Reference Collection.

[136] "Aeronautical Projects Update—June 1977," Memorandum from Director of Aeronautical Projects and Director of Research to the Director of NASA Hugh L. Dryden Flight Research Center, dated July 5, 1977. Part of NASA Dryden Flight Research Center Historical Reference Collection.

[137] Carter, Alan L., *Basic Research Review for the NASA OAST Research Council*, RTOP-743-32-23, Aug. 20, 1974, pp. 43–45.

[138] Edinger, Lester D., Frederick L. Schenk, and Alan R. Curtis, *Study of Load Alleviation and Mode Suppression System (LAMS) on the YF-12 Airplane*, NASA CR-2158, Washington, DC, Dec. 1972, p. 2.

[139] Ibid., pp. 1–5.

[140] "NASA YF-12 Cooperative Control Program" briefing, 1983. Part of NASA Dryden Flight Research Center Historical Reference Collection.

[141] McKay, Jim M., "Request for Project Approval to Measure the YF-12 Structural Response to Aerodynamic Shaker Excitation and to Correlate with Analytical Results," letter to Acting Director NASA Flight Research Center Edwards, CA, Feb. 4, 1975. Part of NASA Dryden Flight Research Center Historical Reference Collection.

[142] Ibid.

[143] Ibid.

[144] Merlin, Peter W., "Lockheed SR-71 Flight Log," Sept. 2002. Part of NASA Dryden Flight Research Center Historical Reference Collection.

[145] Crickmore, *Lockheed Blackbird—Beyond the Secret Missions*.

[146] Merlin, "SR-71 Flight Log."

[147] Crickmore, *Lockheed Blackbird—Beyond the Secret Missions*.

[148] Hallion, Richard P., and Michael H. Gorn, *On the Frontier: Experimental Flight at NASA Dryden*, Smithsonian Books, Washington, DC, 2003, pp. 297–299.

[149] Merlin, "SR-71 Flight Log."

[150] Merlin, Peter W., "Storied SR-71 Is Dryden's Newest Display," *The Dryden X-Press*, Vol. 44 No. 4, Nov. 2002, p. 3.

[151] Miller, *Lockheed Skunk Works: The First Fifty Years*, Aerofax, Arlington, TX, 1993, pp. 210–211.

[152] Ibid.

[153] Aronstein, David C., and Piccirillo, Albert C., *Have Blue and the F-117A: Evolution of the Stealth Fighter*, AIAA, Reston, VA, 1997, pp. 234–236.

[154] Johnson, *History of the Oxcart Program*.

[155] Aronstein and Piccirillo, *Have Blue and the F-117A: Evolution of the Stealth Fighter.*

[156] Merlin, Peter W., "SR-71 Blackbird," *Advanced Processes and Materials*, No. 161(5), ASM International, May 2003, pp. 27–29.

[157] Johnson, Clarence L., "Development of the Lockheed SR-71 Blackbird," *Lockheed Horizons*, No. 9, Winter 1981/82, Lockheed Corp., Burbank, CA, 1981, pp. 2–18.

[158] Goodall and Miller, *Lockheed's SR-71 'Blackbird' Family.*

[159] Johnson, "Development of the Lockheed SR-71 Blackbird."

[160] Goodall and Miller, *Lockheed's SR-71 'Blackbird' Family.*

[161] DeGrey, Richard, Memorandum to Bruce Wright regarding "Problems in Manufacturing, Flight-Testing, and Maintenance of a Mach 2.2 Aircraft vs a Mach 3.0 Aircraft," Lockheed Aircraft Co., Burbank, CA, n.d.

[162] Alitzer, John, "SR-71 Structures and Materials," part of "Case Studies in Engineering: The SR-71 Blackbird," Course Ae107, Presented at the Graduate Aeronautical Lab., California Inst. of Technology, Pasadena, CA, April–May 1990.

[163] Brown, William H., "J58/SR-71 Propulsion Integration or the Great Adventure into the Technical Unknown," *Lockheed Horizons*, No. 9, Winter 1981/82, Lockheed Corp., Burbank, CA, 1981, pp. 7–13.

[164] Johnson, *History of the Oxcart Program.*

[165] Brown, "J58/SR-71 Propulsion Integration or the Great Adventure into the Technical Unknown."

[166] Merlin, Peter W., *Mach 3+: NASA/USAF YF-12 Flight Research, 1969–1979*, NASA SP-2001-4525, NASA Headquarters, Washington, DC, 2002, pp. 87–95.

[167] Ibid.

[168] Hallion, Richard P., *On the Frontier: Flight Research at Dryden, 1946–1981*, NASA SP-4303, Washington, DC, 1984, pp. 186–194.

[169] They derived the model from wind-tunnel test data using a multiple regression technique, which uses a least-squares method to fit equations to the data. The least-squares method is a statistical technique of fitting a curve close to some given points that minimize the sum of the squares of the deviations of the given points from the curve.

[170] Letter from Milton O. Thompson to Walter C. Williams, dated Nov. 1, 1976. Part of NASA Dryden Flight Research Center Historical Reference Collection.

[171] Merlin, *Mach 3+: NASA/USAF YF-12 Flight Research, 1969–1979.*

[172] Matranga, Gene J., "The Realities of Manned Supersonic and Hypersonic Flight— Lessons Learned" presentation at NASA Lewis Research Center Hypersonic Propulsion Symposium Aug. 1988. Part of Matranga's personal collection.

[173] Merlin, *Mach 3+: NASA/USAF YF-12 Flight Research, 1969–1979.*

[174] Merlin, "Storied SR-71 Is Dryden's Newest Display."

INDEX

A-1 project, 10, 11
A-10 project, 16
A-11 concept, 16–17
  CIA reaction to, 17
  radar studies, 17
  YF-12A vs., naming protocol, 40
A-12 concept, 17–19
  acceptance of, 19
A-12 configuration, 19–23
  air-conditioning, 21
  cabin pressure, 20
  cockpit, 20
  drag shute, 21
  electronic compartment, 20–21
  fuel tanks, 22
  fuselage, 19
  landing gear, 21
  mission equipment bay, 21
  nose design, 19–20
  refueling receptacle, 21
  tail design, 22–23
  wing design, 22
A-12 design,
  AF-12 interceptor, 36–42
  Air Force variant, 36
  Article 121, 31
  black paint, 35
  drones, 42–52
  fleet growth, 34
  mission ready, 35
  OXCART, 31–36
  retirement of, 36
  trainer type, 34
A-12 project, 1–2
  OXCART, 1
A-12 systems design, 23–31
  circulating system, 25–26
  cooling, 26

flight control, 26
fuel system, 25
hydraulics, 23–24
pressurization, 26
propulsion, 24–25
A-12, R-12 Universal design vs., 54, 55
A-2 project, 11
A-3 project, 9–10
  engine details, 12
  fuel, 10
A-4 project, 13
A-5 project, 13
A-6 project, 14
A-7 project, 140, 142
A-8 project, 16
A-9 project, 16
Aerodynamics,
  D-21 design and, 48–49
  effect on performance, 87–89
    chines, 88
    delta wings, 87
    differential heating, 88–89
    engine location, 89
    fuel tank sequencing, 88
    stability augmentation system, 87
AF-12 project, 1–3
  crew positions, 38
  external configurations revisions, 38
  Hughes missiles and radar, 37
  interceptor design,
    Article 1001, 37
  KEDLOCK, 1–3, 36–42
  modifications, 37
  YF-12A, 40
AFCS. *See* automatic flight control
  system.
Aft fuselage, 74–75
Afterburner, 104

Air conditioning equipment, 21
Air data display system, 92
Air Force,
  NASA and, joint ventures, 116–118
  SR-71 usage by, 55
  variant, 36
Air induction system, 97
Air temperature, effect on climbing
    performance, 111
Airflow control, 97–102
  inlet system, 99–102
Airframe, powerplant matching with, 160
Alloys, titanium, 62–64
Altitude performance, 105–110
  improving of, 110
  range of, 108–109
  testing of, 105–110
ANS. *See* astroinertial navigation system.
AQUATONE, project, 5–7
Archangel, 1
  A-1 project, 10
  bomber version, 52
    RB-12, 52
  design,
    R-12 Universal, 52–55
    RS-12 reconnaissance/strike
      variant, 54
    SR-71 design, 54–55
  proposals, 8–19
    A-10 project, 16
    A-11 concept, 16–17
    A-12, 17–19
    A-2 project, 11
    A-3, 9–10
    A-4 project, 13
    A-5 project, 13
    A-6 project, 14
    A-7 project, 140, 142
    A-8 project, 16
    A-9 project, 16
    Archangel 1, 10
    Arrow series, 12–13
    G-2A, 10–11
  weaknesses of, 14
Arrow series, 12–13
Article 1001, 37
Article 121, 31
  fuel tank sealants, 31
  maiden flight, 33
  taxi tests, 31–35
Assembly steps, 79–80

Astralloy, 81
Astroinertial navigation system (ANS),
    31, 91
AT-12T, trainer style, 34
AUTO NAV, 28
Automatic flight control system (AFCS),
    27, 91
  astrointertial navigation system, 91
  flight reference system, 91
Automatic navigation (AUTO NAV), 28
Autopilot
  pitch and roll, 92
  system, 27–28
    navigation, 28

B-52H bomber, 50
Batteries, 28
BF Goodrich Silvertown tires, 84
Black paint, 35
Blackbird
  family tree, 1
    A-12, 1
    AF-12, 1–3
    M-21, 1, 3
    numerical designators, 4
    SR-71, 3
    YF-12, 1
  final flights, 144–145
  reactivation of, 142–143
Blackbody radiation, 68
Bleed bypass system, 102
Blip/scan ratio (BSR), 7
Bomber usage, Archangel design and, 52
Boundary layers, 163–164
  measurements, 131
BSR. *See* blip/scan ratio.

Cabin pressure, 20
Canopies, 72–74
CAPCS. *See* cooperative airframe/
    propulsion control system.
Central airborne performance analyzer
    (CAPA), 123–124
Central Intelligence Agency, 1
  Lockheed, cooperation between, 58
Chines, 88
Chordwise corrugations, titanium and, 59
CIA
  Lockheed and, 17
  reaction to A-11 concept, 17
Circulating system, auxiliary, 25–26

Climbing performance, air temperature
    effect on, 111
Cockpit, 27
    controls, 27
Coldwall experiment, 132–134
Combustion section, 103
Communications, 29–30
    high-frequency radio, 30
    low-frequency radio, 30
    tactical air navigation, 30
Computer modeling, lack of, 156
Contaminants, titanium alloys and, 62–63
Convair proposal, 7–8
    FISH, 7
    KINGFISH, 7–8
Cooling systems, 26
Cooperative airframe/propulson control
    system (CAPCS), 137
Crew revisions, AF-12 interceptor design
    and, 38
Customer relations, 150

D-21 design, 44
    aerodynamic problems, 48–49
    cancellation of, 51–52
    engine, 47–48
    failures of, 50
    fuselage, 44–45
    magnesium-thorium alloy, 48
    materials used, 44
    mockup, 48
    modification to, 50
        B-52H bomber, 50
    nose-cone, 48
    production of, 49
    size, 44
    tail, 47
    trials, 50
    wing structure, 46–47
DAFICS. *See* digital automatic flight and
    inlet ground system.
Data collection, NASA and, 123–124
Delta wings, 87
Design and production principles, 147–154
Design evolution, 5–55
    GUSTO, 7
    proposals, 7
        Archangel, 8–19
        Convair, 7–8
        SUNTAN program, 6
        U-2 reconnaisance airplane, 5–7

DFBW. *See* digital fly by wire.
Differential heating, effects on
    aerodynamics, 88–89
Diffusers, 103
Digital automatic flight and inlet
    ground system (DAFICS),
    82, 162
Digital fly by wire (DFBW) aircraft, 129
Drag penalty determination, 131–132
Drag shute location, 21
Driveshafts, 97
Drone aircraft, 42–52
    D-21, 44
    M-21, 43
    McNamara, Robert, 43
    MD-21, 44
    naming of, 43–44
    NASA and, 130–131
    TAGBOARD, 42–52
Dual mode landing gear system,
    126–128

Ejection seat, 28–29
Electrical system, 28
Electronic compartment, 20–21
Elevons, 77
Elgiloy, 81
Engines, 80–82
    airflow issues, 157
    Astralloy, 81
    cooling issues, 157
    design, D-21 and, 47–48
    digital automatic flight and inlet ground
        system, 82
    flight controls, 81–82
    Hastelloy-X, 81
    inlet spike assembly, 79
    location of, 89
    Pratt and Whitney JTD-11B-20, 80–81
    problems with, 158–159
    starting, 104
    Waspalloy, 80
Exhaust gas temperature sensors, 102

FATOLA. *See* Flexible Aircraft Takeoff
    and Landing Analysis.
Fiberglass, 65
Final flights, 144–145
FISH proposal, 7, 12
Flexible Aircraft Takeoff and Landing
    Analysis (FATOLA), 125–127

Flight control systems, 26, 81–82, 89–92
  air-data display system, 92
  astroinertial navigation system, 31
  automatic, 27
  autopilot, 27–28
  batteries, 28
  cockpit controls, 27
  communications, 29–30
  ejection seat, 28–29
  electrical, 28
  flight instruments, 30–31
  identification, friend or foe, 31
  life support, 29
  Mach trim system, 28, 92
  navigation, 30
  pitch and roll autopilot, 92
  primary, 89–90
  radar beacon, 31
  recorder, 31
  reference, 30
Flight instruments, 30–31
Flight range, 111–113
  pilot's effect on, 113
Flight reference system (FRS), 91
Flight testing, 151
Flying wind tunnel, 165
Forward fuselage, 72
FRS. *SEE* flight reference system.
Fuel, 10, 82–83
  air temperature effect on, 111
  consumption, 110
  innovation of, 155, 158
  system, 25, 83
  tanks, 22
    sealants, 31
    seating, 83
    sequencing, 88
    triethylborane, 82
Funding, 58
Fuselage, 19
  design, D-21, 44–45

G-2A project, 10–11
  RCS issues, 11
Gearboxes, problems with, 158
Grease, 83
GUSTO project, 7

Handling qualities, 89
  nose-up trim, 89

research, 163
trim drag, 89
Hastelloy-X nickel alloy, 70, 81
Heating measurements, 161
High Temperature Loads Laboratory, 120–121
High temperature radar-absorbing materials, innovative use of, 154–155
High temperature window glass, innovation of, 155
High-frequency radio, 30
Historical timeline, 177–182
Hughes Aircraft
  missiles, 37
  radar, 37
Human behavior, 163
Hydraulic oil, 83
Hydraulics, 23–24
  landing gear, 23–24

Identification, friend or foe (IFF) transponder, 31
IFF transponder, 31
Inlet control, DAFICS, 162
Inlet guide vanes, 102
Inlet spike assembly, engine, 79
Inlet system control, 99–102
  bleed bypass system, 102
  inlet guide vanes, 102
  shock waves, 101–102
Instrumentation research, NASA and, 140
Instruments, flight, 30–31
IRIDIUM system, 142

Johnson, Clarence L. "Kelly," 1, 12, 13, 14, 36, 37, 38, 40, 43, 50, 52, 57–59, 130–131
  design and production principles, 147–154
  observations re NASA, 118
Johnson, Lyndon , 40, 54, 105

KEDLOCK project, 1–3, 36–42
KINGFISH proposal, 7–8
Kirchoff's Law of Radiation, 67

LAMS. *See* loads alleviation and mode suppression system.

Landing gear, 21, 23–24, 84–86
  dual mode, 126–128
  load factors, 86
  main truck, 84–85
  nose gear, 86
  tires, 84
  wheel assemblies, 86
Landing studies, 124–129, 162–163
  low aspect ratio supersonic aircraft, 125–127
  space shuttle landing dynamics, 124–125
  total in-flight simulator, 129
LASRE. *See* Linear Aerospike SR-71 experiment.
Life support system, 29
  liquid oxygen, 29
Linear Aerospike SR-71 experiment (LASRE), 143–144
Liquid oxygen (LOX) system 29
Load factors, landing gear and, 86
Loads alleviation and mode suppression (LAMS) system, 136–137
Loads suppression, 164
Lockalloy, 75
  skin panels, 75, 134–135
Lockheed, 17, 38
  acceptance of A-12 concept, 19
  Central Intelligence Agency, cooperation between, 58
  contemporary issues, 153–154
  design and production principles, 147–154
Low aspect ratio supersonic aircraft, 125–127
  dual mode landing gear system, 126–128
  Flexible Aircraft Takeoff and Landing Analysis, 125–127
LOX system. *See* liquid oxygen.
Lubricants, innovation of, 155, 158

M-21 design and project, 1, 3, 43
Mach speed cruising, 95
Mach trim system, 28, 92
Magnesium-thorium (magthor) alloy, 48
Magthor alloy, 48
Maiden flight, 33
Main gear truck, 84–85
Manufacturing,
  management of, 151
  techniques, 64

Materials, 65–70
  nickel alloys, 69–70
  paint, 66–67
  plastic laminates, 65–66, 70
  fiberglass, 65
  phenyl silane, 65, 70
  silicone-asbestos, 65, 70
  steel, 69
  titanium alloys, 65, 68
McNamara, Robert, 41, 43
Missile systems, AF-12 interceptor and, 37
Mission equipment bay, 21
Mode alleviation, 164
Monel alloy fasteners, 68
Monocoque structure, 70
Multispar/multirib wing, 70–71

Nacelle structure, 77–79
NASA, 115–145
  Air Force and, joint ventures, 116–118
  Blackbird reactivation, 142–143
  High Temperature Loads Laboratory, 120–121
  Johnson, observations re, 118
  landing studies, 124–129
  obtaining Blackbird, 122
  propulsion research, 121–124
    central airborne performance analyzer, 123–124
    data collection, 123–124
    unstarts, 124
    wind-tunnel data points, 124
  research laboratories, 129–139
    boundary-layer measurements, 131
    Coldwall experiment, 132–134
    cooperative airframe/propulsion control system digital computer, 137
    drag penalty determination, 131–132
    drones, 130–131
    instrumentation, 140
    IRIDIUM system, 142
    Linear Aerospike SR-71 experiment, 143–144
    loads alleviation and mode suppression system, 136–137
    Lockalloy skin panels, 134–135
    optical air-data system, 141
    ozone layer experiments, 142

NASA (*Continued*)
  remote-sensing technology, 140–141
  shaker vane system, 136
  sonic booms, 142–143
  structural analysis, 138–139
  structural panels, 135
  subsonic wake vortex flow, 134
  SR-71 usage by, 55
NASA, wholly owned Blackbirds, 139–142
NASTRAN program, 138–139
Navigation system, 30
Nickel alloys, 69–70
  hastelloy-X, 70
  Rene' 41, 69–70
Nose
  cone design, D-21 and, 48
  gear, 86
  structure, 71–72
  up trim, 89
Nose design, 19–20

OADS. *See* optical air-data system.
Oil, hydraulic, 83
Operating characteristics, 92–95
  affecting factors, 94–95
  Mach speed cruising, 95
  pilot involvement in, 95
  supersonic climb, 94
  transonic acceleration, 93–94
Optical air-data system (OADS), 141
OXCART project, 1, 31–36
Ozone layer experiments, 142

Paint, 66–67
  blackbody radiation, 68
  Kirchoff's Law of Radiation, 67
Payload capacity, 113
Performance, 87–113
  aerodynamics, 87–89
  altitude and speed, 105–110
  altitude and speed, testing of, 105–110
  automatic flight control system, 91
  climbing and cruising, 110–111
  flight control system, 89–92
  flight range, 111–113
  fuel consumption, 110
  handling qualities, 89
  operating characteristics, 92–95
  payload capacity, 113
  propulsion system, 95–105

PFCS. *See* primary flight control system.
Phenyl silane, 65, 70
Pickling methods, titanium and, 60
Pilot,
  flight range and effect on, 113
  workload, operating characteristics and, 95
Pistons, 83
Pitch and roll autopilot, 92
Plastic laminates, 65–66, 70
  fiberglass, 65
  phenyl silane, 65, 70
  silicone-asbestos, 65, 70
Powerplant, airframe matching with, 160
Pratt and Whitney,
  JT11D-20 (J58) engines, 95–97
    driveshafts, 97
    problems with, 96–97
  JTD-11B-20, 80–81
President Lyndon Johnson, 54
Pressurization systems, 26
Primary flight control system, (PFCS), 89–90
  stick motions, 90
Production summary, 167–176
Project AQUATONE, 5–7
Project GUSTO, 7
Project management, 149–150
  contemporary issues, 153–154
  flight testing, 151
  manufacturing, 151
  relations with customer, 150
  specifications, 150–151
Propulsion data, 162
Propulsion design, 24–25
Propulsion, NASA and, 121–124
  central airborne performance analyzer, 123–124
Propulsion research, NASA and
  data collection, 123–124
  unstarts, 124
  wind-tunnel data points, 124
Propulsion system, 95–105
  afterburner, 104
  air induction system, 97
  airflow control, 97–102
  combustion section, 103
  diffusers, 103
  engine start, 104

exhaust gas temperature sensors, 102
Pratt and Whitney JT11D-20 (J58)
engines, 95–97

Quality assurance, 151
Quality control, structural materials and, 61

R-12 Universal design, 52–55
　A-12 vs., 54, 55
　configuration of, 54
Radar beacon system, 31
Radar cross section (RCS), 6
　G-2A project and, 11
Radar studies, A-11 concept and, 17
Radar systems, AF-12 interceptor and, 37
Radar tracking
　blip/scan ratio (BSR), 7
　radar cross section (RCS), 6
　U-2, 6
RB-12, 52
RCS. *See* radar cross section.
Recorder system, 31
Refueling receptacle, 21
Remote sensing, 165
　technology experiments, 140–141
Rene' 41 nickel alloy, 69–70
Research laboratories, 129–139
　results of, 160–166
　　boundary layer, 163–164
　　flying wind tunnel, 165
　　handling-qualities, 163
　　heating measurements, 161
　　human factors, 163
　　inlet control, 162
　　landing studies, 162–163
　　loads suppression, 164
　　mode alleviation, 164
　　propulsion data, 162
　　remote-sensing, 165
　　upper atmosphere physics, 164
　　wind-tunnel modeling, 162
Retirement, A-12 design, 36
Rich, Ben, 152–153
Rivets, 61–62
RS-12 reconnnaissance/strike variant, 54

SAS. *See* stability augmentation system.
Sealants,
　fuel tank, 31
　innovation of, 155, 158

Seals, 83
Security requirements, 151–152
SENIOR CROWN project, 3
Shaker vane system, 136, 164
Shock waves, inlet system control and 101–102
Shuttle landings,
　digital fly by wire aircraft, 129
　dynamics of, 124–125
Silicone-asbestos, 65, 70
Sonic booms, NASA research on, 142–143
Specifications, 150–151
Speed performance, 105–110
　improving of, 109–110
　maximum levels, 107–108
　testing of, 105–110
Spot welding, 69
SR-71 astroinertial navigation system, 31
SR-71 design
　Air Force usage of, 55
　NASA usage of, 55
　use of, 55
SR-71 project, 3
　SENIOR CROWN, 3
Stability augmentation system (SAS), 87, 91
Stainless steel honeycomb structure, 58
Standoff clip, 16
Static test articles, 156
Steel, 58
　corrosion resistant, 69
　stainless steel honeycomb structure, 58
　honeycomb structure, 58
Stick motions, 90
Structural analysis program, 138–139
Structural materials, 58–64
　quality control, 61
　steel, 58
　titanium, 58
Structural panel materials,
　experiments re, 135
Structure,
　aft fuselage, 74–75
　canopies, 72–74
　elevons, 77
　engine inlet spike assembly, 79
　forward fuselage, 72
　monocoque, 70
　multispar/multirib wing, 70–71

Structure (*Continued*)
  nacelle, 77–79
  nose, 71–72
  tail, 77
  wing assemblies, 75–77
Subsonic wake vortex flow experiments, 134
SUNTAN progam, 6
Supersonic aircraft, low aspect ratio, 125–127
Supersonic climb, 94

Tactical air navigation (TACAN), 30
TAGBOARD, 42–52
Tail design, 22–23
  D-21 and, 47
Tail structure, 77
Taxi tests, 31–35
TEB. *See* Triethylborane.
Technology gains, 154–160
Technology innovation,
  advanced computer modeling, 156
  engine problems, 158–159
  fuels, 155, 158
  engines, 157–58
    airflow, 157
    cooling issues, 157
  gearbox problems, 158
  high temperature
    radar-absorbing materials, 154–155
    window glass, 155
  lubricants, 155, 158
  power plant to airframe matching, 160
  sealants, 155, 158
  standoff clip, 156
  static test articles, 156
  titanium, 154
  unstart, 159
TIFS. *See* total in-flight simulator.
Timeline, 177–182
Tires, 84
Titanium, 58
  alloys, 62–64, 65, 68
    contaminants, 62–63
    tool cuttings, 63–64
  chordwise corrugations, 59
  cockpit tests, 59–60
  innovative use of, 154
  manufacturing techniques, 64
  material tests, 58–69
  pickling methods, 60
  rivets, 61–62, 68–69
    Monel alloy fasteners, 68
  spot welding, 69
Tool cuttings, titanium alloys and, 63–64
Total in-flight simulator (TIFS), 129
Transonic acceleration, 93–94
Triethylborane (TEB), 82
Trim drag, 89

U-2 reconnaisance airplane, 1, 5–7
  life span of, 5–6
  Project AQUATONE, 5–7
  radar tracking of, 6
Unstarts, 124
  problems, 159
Upper atmosphere physics, study of, 164

Waspalloy, 80
Wheel assemblies, 86
Wind tunnel, 165
  data points, 124
  modeling, 162
Window glass, innovation of, 155
Wing,
  assemblies, 75–77
  design, 22
  multispar/multirib, 70–71
  structure, D-21 design and, 46–47

Y-12A design, cancellation of, 41, 42
  McNamara, Robert, 41
YF-12 project, 1
YF-12A design, 40
  A-11, naming protocol, 40
  maiden flight, 40

SUPPORTING MATERIALS

Many of the topics introduced in this book are discussed in more detail in other AIAA publications. For a complete listing of titles in the AIAA Library of Flight series, as well as other AIAA publications, please visit http://www.aiaa.org.

The accompanying disc contains the following supporting materials.

SR-71 AND YF-12 RELATED MANUALS, REPORTS, AND IMAGES

The following manuals, reports, photographs, and drawings are included on the disc; file names for images are listed in bold type.

MANUALS AND REPORTS

A-12 Flight Manual
Archangel to Oxcart
Comments on ADP Operation
Design Study Archangel Aircraft
Lockheed SR-71 Handbook
Manufacturer's Model Specification
Proposal for Lightweight Reconnaissance Aircraft
SR-71 Flight Manual
YF-12 Flight Manual

KEDLOCK IMAGES

The original AF-12 forward fuselage mock-up closely resembled the A-12 on which it was based. The prominent radome and infrared sensors had not yet been incorporated into the design. (Courtesy of Lockheed Martin Corp.) (**KED_1**)

The second YF-12A, as it appeared in 1963, was only painted black along its wing edges, chines, and radome. (Courtesy of Lockheed Martin Corp.) (**KED_2**)

A front view of the second YF-12A shows the cut-off chines just aft of the radome. (Courtesy of Lockheed Martin Corp.) (**KED_3**)

This profile view of the first YF-12A illustrates a variety of features including the ventral fin and strakes, infrared sensor on the nose, and a nacelle-mounted camera pod used to photograph missile separation tests. (Courtesy of Lockheed Martin Corp.) (**KED_4**)

The third YF-12A set several world speed and altitude records on 1 May 1965. Here, it appears in the all-black paint scheme m adopted in 1964 to better radiate heat from the aircraft's skin. (Courtesy of Lockheed Martin Corp.) (**KED_5**)

The AF-12 (later designated YF-12A) fire control system and weapon bays occupied the space normally reserved for reconnaissance equipment on other Blackbird variants. (Courtesy of Lockheed Martin via Gene Matranga) (**KED_6**)

This shadow of the second YF-12A dramatically illustrates the lack of chines on the aircraft's radome. (Courtesy of NASA) (**KED_7**)

This close up shows one of two shaker vanes installed on the YF-12A. Data from these flights were compared with data generated by the NASA Structural Analysis (NASTRAN) computer program. (Courtesy of NASA) (**KED_8**)

In the forward cockpit (pilot's station) of the YF-12A, vertical strip displays replaced some of the round-faced instruments found in the A-12 and SR-71. (Courtesy of Lockheed Martin Corp.) (**KED_9**)

The aft cockpit of the YF-12A accommodated the fire control officer during Air Force missile test missions, or a flight test engineer for NASA research flights. (Courtesy of Lockheed Martin Corp.) (**KED_10**)

OXCART IMAGES

In March 1959, Lockheed technicians treated a model of the A-10 design with antiradar coatings and tested it inside the company's anechoic chamber. The radar-absorbent panels were added to the model without regard for aerodynamic issues. (Courtesy of Lockheed Martin Corp.) (**OXC_1**)

The first A-12 is seen here during final assembly. The all-metal chines and edges are clearly visible. (Courtesy of Lockheed Martin Corp.) (**OXC_2**)

Prior to flight, the first A-12 underwent extensive ground testing. Initial engine runs were conducted using external fuel tanks mounted atop the wings because fuel tank sealant for the wing interior had yet to be perfected. (Courtesy of Lockheed Martin Corp.) (**OXC_3**)

The first A-12 made its initial flights in an unpainted state. The natural metal finish provides a clear view of the edge structure. (Courtesy of Lockheed Martin Corp.) (**OXC_4**)

Because the J58 engines were not yet available, the first A-12 made its early flights powered by J75 engines. During these flights, the aircraft's inlet spikes were fixed in place. (Courtesy of Lockheed Martin via Jim Goodall) (**OXC_5**)

Here the first A-12 dumps fuel during a test flight. The F-101 aircraft, at right, provided safety chase. (Courtesy of Lockheed Martin via Jim Goodall) (**OXC_6**)

The second A-12 initially served as a pole model for radar cross-section measurements. It was mounted, inverted, on a specially designed pylon. (Courtesy of Lockheed Martin Corp.) (**OXC_7**)

The second A-12 is seen here taking on fuel from a KC-135Q tanker. Aerial refueling extended the operational range of the A-12. (Courtesy of Lockheed Martin via Jim Goodall) (**OXC_8**)

The A-12T was the sole trainer variant prior to production of the SR-71B and SR-71C. It had an instructor's cockpit located above and behind the student cockpit. The A-12T was powered by J75 engines. (Courtesy of Lockheed Martin via Jim Goodall) (**OXC_9**)

This was the first A-12 photo released when the airplane was declassified in the early 1980s. (Courtesy of Lockheed Martin Corp.) (**OXC_10**)

SENIOR CROWN IMAGES

Lockheed technicians construct Blackbirds on the company's SR-71 assembly line in Burbank. (Courtesy of Lockheed Martin Corp.) (**SEN_1**)

The first SR-71 awaits its maiden flight at Air Force Plant 42 in Palmdale. (Courtesy of Lockheed Martin Corp.) (**SEN_2**)

A Lockheed F-104 provided safety chase during the maiden flight of the SR-71. (Courtesy of Lockheed Martin Corp.) (**SEN_3**)

During high-speed cruise, the SR-71 experienced surface temperatures between 400 and 1,200°F. (Courtesy of Lockheed Martin Corp.) (**SEN_4**)

Plastic chine and edge panels are visible due to the difference in surface contrast against the titanium fuselage. (Courtesy of Lockheed Martin Corp.) (**SEN_5**)

As the SR-71B lands, a drag chute deploys to air deceleration. (Courtesy of Lockheed Martin Corp.) (**SEN_6**)

The Blackbird's high-speed landing characteristics caused tremendous wear on the tires. (Courtesy of Lockheed Martin Corp.) (**SEN_7**)

The SR-71A banks steeply, shortly after takeoff. (Courtesy of Lockheed Martin Corp.) (**SEN_8**)

The internal arrangement of the SR-71 allowed for no wasted space. (Courtesy of U.S. Air Force) (**SEN_9**)

During the Linear Aerospike SR-71 Experiment, the airplane carried a scale half-span model of the X-33 with a functional aerospike engine. (Courtesy of NASA) (**SEN_10**)

TAGBOARD IMAGES

The mated combination of the M-21 mothership and D-21 drone was called the MD-21. (Courtesy of Lockheed Martin Corp.) (**TAG_1**)

The drone's inlet and exhaust were covered with aerodynamic fairings during captive test flights. (Courtesy of Lockheed Martin Corp.) (**TAG_2**)

A dorsal pylon supported the D-21 atop the M-21 mothership. (Courtesy of Lockheed Martin via Jim Goodall) (**TAG_3**)

This close-up view shows the aerodynamic inlet fairing in detail. The frangible plastic faring could be separated in flight to allow for engine start. (Courtesy of Lockheed Martin Corp.) (**TAG_4**)

Separation of the D-21 from the M-21 created complex aerodynamic interference fields between the two vehicles. (Courtesy of Lockheed Martin via Gene Matranga) (**TAG_5**)

Technicians prepare to mate the first D-21 to the M-21 mothership. (Courtesy of Roadrunners Internationale) (**TAG_6**)

The B-52H carried two D-21B drones and their rocket boosters on wing pylons. (Courtesy of Lockheed Martin via Jim Goodall) (**TAG_7**)

While on the ground, the booster's ventral fin was folded to one side to allow ground clearance. A ram-air turbine on the booster's nose provided the D-21B with electrical and hydraulic power prior to launch. (Courtesy of Lockheed Martin via Jim Goodall) (**TAG_8**)

Many parts of the D-21 and D-21B drones were made from composite plastics. (Courtesy of Lockheed Martin Corp.) (**TAG_9**)

This diagram shows the sequence of events involved in a D-21B mission. (Courtesy of U.S. Air Force) (**TAG_10**)

HISTORY AND MISSION VIDEOS OF THE SR-71

A brief history of the YF-12 and SR-71 at Dryden Flight Research Center is depicted in the first video on the disc. The second video shows flight preparations, taxi, takeoff, and the SR-71 in flight. A sequence of scenes from several missions is included in the third video, comprising pilot and engine preparations, taxi, takeoff, mission highlights, and landing.